ISNM
International Series of Numerical Mathematics
Vol. 128

Pattern Formation in Viscous Flows

The Taylor-Couette Problem and Rayleigh-Bénard Convection

Rita Meyer-Spasche

Birkhäuser Verlag
Basel · Boston · Berlin

Author:

Rita Meyer-Spasche
Max-Planck-Institut für Plasmaphysik
D-85748 Garching bei München
Germany

1991 Mathematics Subject Classification 35Qxx, 65Nxx, 76Exx

A CIP catalogue record for this book is available from the Library of Congress, Washington D.C., USA

Deutsche Bibliothek Cataloging-in-Publication Data

Meyer-Spasche, Rita:
Pattern formation in viscous flows : the Taylor-Couette problem and Rayleigh-Bénard convection / Rita Meyer-Spasche. – Basel ; Boston ; Berlin : Birkhäuser, 1999
(International series of numerical mathematics ; Vol. 128)
ISBN 3-7643-6047-X (Basel ...)
ISBN 0-8176-6047-X (Boston)

Printed on acid-free paper produced of chlorine-free pulp. TCF ∞
Cover design: Heinz Hiltbrunner, Basel

ISBN 3-7643-6047-X
ISBN 0-8176-6047-X

9 8 7 6 5 4 3 2 1

Preface

> It seems doubtful whether we can expect to understand fully the instability of fluid flow without obtaining a mathematical representation of the motion of a fluid in some particular case in which instability can actually be observed, so that a detailed comparison can be made between the results of analysis and those of experiment.
>
> – G.I. Taylor (1923)

Though the equations of fluid dynamics are quite complicated, there are configurations which allow simple flow patterns as stationary solutions (e.g. flows between parallel plates or between rotating cylinders). These *flow patterns* can be obtained *only in certain parameter regimes.* For parameter values not in these regimes they cannot be obtained, mainly for two different reasons:

- The mathematical *existence* of the solutions is parameter dependent; or
- the solutions exist mathematically, but they are not *stable.*

For finding stable steady states, two steps are required: the steady states have to be found and their stability has to be determined.

Only a few laminar flows correspond to explicitly known solutions of the equations of motion. In the case of the Taylor problem, for instance, only Couette flow (1.12) is explicitly known. Even fewer flows than explicitly known are simple enough to allow the detailed analysis of their stability with the methods of mathematical physics. The recent development of large computers and of numerical methods makes it possible to solve the full nonlinear equations for many parameter values and to investigate the stability properties of the computed steady states. Scientific computing created completely new possibilities. Nevertheless, only all available methods, i.e.

- experiments with fluids;
- explicit solution of the equations, when possible, and perturbation analysis in a neighborhood of such solutions;
- numerical solution of the full nonlinear equations and numerical stability analysis

together give a more or less complete picture. In the following text we will see again and again how results on the Taylor problem obtained with different methods complement and supplement each other.

It is one of the merits of G.I. Taylor that he saw which aspects of flows between rotating cylinders can or must be neglected to obtain a simplified model which is accessible to mathematical analysis. In his work of 1923 he obtained calculated as well as measured results – impressive for their vastness, completeness and accuracy. Even more impressive are the comparison and good agreement (less than 5% deviation of the calculated from the measured values, with about 2% errors in the measurements).

Since then, the '*Taylor problem*' is a popular research subject, because it is so simple and at the same time so complex. The investigations discussed in the following were performed by mathematicians, physicists and engineers. Contact and exchange between disciplines is typical for work on the Taylor problem: from the beginning, results obtained by theoretical investigations and by experiments were compared to each other, and this comparison was always considered important [Tay, AH92]. Often the Taylor problem was a test case for newly developed methods, and it initiated and stimulated the development of new methods.

For *mathematicians* it is one of the popular examples, especially:

- for studying bifurcations (pitchfork, Hopf and homoclinic), hysteresis and catastrophes; and
- for the development and testing of new methods (analytical and numerical).

Because of the successful cooperation with other disciplines it is a good example of *Applied* Mathematics.

Physicists use it especially for studying:

- the laminar – turbulent transition and
- the occurrence of instabilities owing to the dynamical effects of rotation or (more basically) the occurrence of instabilities of fluid flow.

Engineers are especially interested because of *practical applications*:

- In bearings, an axle or pin rotates in a liquid. This liquid is confined or flows in axial direction. Because of the rotating axle it is practically impossible to keep such a configuration perfectly closed to vacuum. This is why the bearings of satellites, of space shuttle etc. do not simply contain oil, but a 'magnetic suspension': a mixture of plastic-coated iron particles and oil. A magnetic field then confines the iron particles, and surface effects force the oil to stay with the particles [Stie].

- In turbines there is a roughly similar situation, but a blade turns and the flow through the configuration is important.
- For separation of bio fluids, e.g. for separation of liquid and corpuscles of blood, stationary Taylor vortices with radial through flow have been used: the liquid passes through the porous outer cylinder, the corpuscles stay in the 'sieve'. The vortex movement cleans the outer cylinder and keeps it from filling up. In this process the blood corpuscles do not get damaged as much as in the older centrifugal process.
- Also, stationary Taylor vortices with radial through flow and a porous outer cylinder have been used for washing wool.
- During chemical reaction processes it is important that the flows passing each other have well-defined speeds, in order that the reacting fluids mix in well-defined mass ratios. The control of speed via control of Reynolds number is relatively simple. In the Taylor apparatus it is achieved by control of the angular speeds of the cylinders. The mixing of the reacting fluids happens at the boundaries of the wavy vortices: the local velocity changes considerably at these interfaces and this is favorable for mixing. In the interior of the vortices there is not much mixing – the diffusion into the interior of the vortices is too slow [SKC].

In the following we will consider only the simplest basic configuration of the Taylor problem: a Newtonian fluid with constant density, viscosity and temperature between concentric cylinders with periodic boundary conditions in the axial direction. There is a considerable amount of research on variations of this basic configuration [AH92, Roe85, Hol83]:

- short cylinders with rigid lids on top and bottom (boundary effects are not neglected): the Benjamin problem;
- variation of the geometric configuration: concentric spheres or ellipses, non-concentric cylinders, variation of gap width with length, etc.;
- additional through flow in the radial or axial direction;
- additional temperature gradient in the radial direction;
- variation of the fluid: non-Newtonian fluids like magnetic suspensions, polymer solutions, liquid crystals, etc.

The earth and also other planets rotate. Vortex rings are sometimes visible in their atmosphere (rising air over the tropic rain forests, sinking air over the desert belt of the earth). In this case there is an additional temperature gradient which must not be neglected. For studying meteorological questions another model problem is thus relevant, the *Bénard problem*: a fluid in a gravitational field between parallel plates or spheres of different temperatures. According to more recent insights, the compressibility of the atmosphere is more important than its viscosity [Lo93a]. This reduced the importance of the Bénard problem for meteorology. But it is still important for the investigation of flows in other geophysical fluids (in the oceans, in the liquid core of the earth etc.) and also in astrophysics. Besides this, it is a very important model problem for the same basic questions for which the Taylor problem is studied.

There is an abundant amount of literature on both problems. Here only books and review articles are listed:
Taylor problem: [Cha, DR, DPS, Stu, Wim, BC86, AH92, Do92, CI94];
Bénard problem: [Cha, DR, Bus85, Bus89, Strau].

The reviews on the Taylor problem mostly concern experiments and results obtained with perturbation methods and with methods of mathematical physics. The rich literature on numerical investigations consists of articles in journals and in proceedings volumes. They are not adequately mentioned in general reviews. To my knowledge there is no general review that adequately integrates those results which were obtained by numerical methods.

In the following text, we will discuss the Taylor problem in great detail. When discussing the Rayleigh-Bénard problem we will concentrate on those aspects which are of importance for understanding the Taylor problem.

I am very grateful to all who gave me the opportunity for joint work and/or intensive discussion on the topics of this book. Special thanks are due to *Herb Keller* and *Philip Saffman*, who introduced me to the Taylor problem, to *Fritz Busse*, who taught me that I should look at the Rayleigh-Bénard problem to understand the Taylor problem better, to *Dietrich Lortz*, to *Frank Pohl*, who did the programming for the (until now unpublished) investigations of the 9×9 model in Chapter 4, to *Eva Sombach*, who made many figures look better, and to *John Bolstad*, who critically commented on the whole manuscript.

Contents

Chapter 1

The Taylor Experiment

In this chapter we discuss the Taylor experiment and mathematical modeling of it. We review experimental results and investigations of the stability of Couette flow. Numerical modeling of Taylor vortex flows will be discussed in the next chapter.

1.1 Modeling of the experiment

1.1.1 Introduction

In homogeneous fluids instability often is caused by dynamic effects of rotation. An important example for this type of centrifugal instability is flow between rotating cylinders. The instability of rotating fluid was first investigated by **Lord Rayleigh** (1880), in the case of **inviscid fluids** ($\nu = 0$).

Though the assumption $\nu = 0$ seemed to be a straightforward simplification in physics, it has strong consequences for the mathematical nature of the describing equations: they are then the (historically older) Euler fluid equations instead of the Navier-Stokes equations. For $\nu = 0$ the second derivatives in (1.3)–(1.6) are not present, and thus part of the boundary conditions for the full system cannot be satisfied. If the problem is further simplified by considering only circumferential flows $(0, v(r), 0, p(r))$, r being the distance from the axis of rotation, flows with arbitrary circumferential velocities $v(r) = r\Omega(r)$ are possible.

Heuristic physical reasoning led Rayleigh to derive his famous stability criterion, see for instance [DR, p. 71ff]. He also noticed an analogy to gravitational instability of stratified fluids at rest (⇒ Rayleigh-Bénard convection, see below).

Rayleigh's Circulation Criterion: *Necessary and sufficient for stability of an axisymmetric flow of angular velocity $\Omega(r)$ with respect to axisymmetric disturbances is:*

$$\Phi(r) := \frac{1}{r^3}\frac{d}{dr}\Big(r^2\Omega(r)\Big)^2 \geq 0. \tag{1.1}$$

If a flow is called stable if the kinetic energy of axisymmetric, z-periodic perturbations is bounded for all admissible initial conditions, this criterion can be proved mathematically [Cha, §67], [Lo93b].

The stability of viscous ($\nu \neq 0$) rotating fluids was first investigated by chance. In 1881, **M. Margules** [Marg] in Vienna suggested to measure the viscosity ν of a fluid by putting it into the gap between two concentric cylinders, rotating one of them, and measuring the torque exerted onto the other cylinder. **Couette** (1890) measured the viscosity of water of different temperatures by this technique and found values close to the values known today. For small angular velocities he observed the expected linear relation. For larger angular velocities he observed a different inclination. This was interpreted as onset of turbulence (see Figs. 1.2 and 1.3 below and eqs. (1.14) and (2.86)).

Mallock (1888, 1896) independently performed similar experiments. Though he also focused on measuring viscosity, **Lord Kelvin** used his experiment to get insights on the stability of viscous flows between rotating cylinders (letter to Rayleigh [Do92]). Later on **G.I. Taylor** looked at Mallock's experiments under the viewpoint of stability [Tay]. He found that Mallock's results partially contradicted Rayleigh's criterion (1.1) for inviscid flows and that it seemed practically impossible to deduce any rules from these experiments. Taylor analyzed and criticized Mallock's experiments very carefully and pointed out the following possible sources of error:

a) length and diameter of the outer cylinder were nearly of the same size (ca 20 cm);
b) Mallock used the full lengths of the cylinders for his measurements;
c) one of the two cylinders was probably not held rigidly enough and could perform small lateral movements.

Mallock attempted to construct his three apparati such that 'the water' in the gap 'is very nearly in the same condition it would be if' the two cylinders 'were infinitely long' [Do92]. He substituted the original bottom by one of liquid mercury in the course of his work. This suggests that he himself noticed that influences of the bottom were a problem. Taylor guessed that a) and b) led to the result that Mallock's experiment featured instability by 3-dimensional disturbances favored by the bottom and was thus dominated by end effects [Tay]. He found new approaches such that these problems were carefully avoided and such that his experiment could actually be modeled mathematically with periodic boundary conditions in infinitely long cylinders. Also, he constructed his apparatus such that both cylinders could rotate independently. Thus a wider parameter regime became accessible.

In dimensional variables, the problem has the following quantities to be chosen by the experimenter:

R_1 radius of the inner cylinder;
R_2 radius of the outer cylinder;
H total length of the cylinders;
L_o length of that portion of the cylinders where the measurements are made;
Ω_1 angular velocity of the inner cylinder;
Ω_2 angular velocity of the outer cylinder;
ν kinematic viscosity of the fluid.

Experimenters usually give the **aspect ratio**

$$\Gamma := H/D, \quad D := R_2 - R_1 \tag{1.2}$$

of their experiment.

Following Taylor, the Taylor problem in the strong sense has theoretically **infinitely long cylinders**, and the experimenters using Taylor's model usually made sure that the published results do not depend on the aspect ratio. To achieve negligibly small end effects,

- the aspect ratio Γ is made large (Taylor used $\Gamma \geq 90$);
- measurements are performed only in the middle portion of the cylinders, far away from top and bottom (Taylor used 20 cm out of 90 cm);
- the lids at top and bottom are designed in a sophisticated way.

Today, the most advanced way to mimic infinitely long cylinders is using **ramps** at the end portions of the cylinders: the radius of the inner cylinder may be increased or the radius of the outer cylinder may be decreased towards the ends, with optimized length and inclination angle [AC, RP87].

Also, the influence of other effects which are not included in the mathematical model but which could be found in a real experiment must be kept small enough to be negligible. They are of two types: simplifying idealizations and unavoidable imperfections. Idealizations typically are eliminated when science gets more advanced and they turn out to be too unrealistic; imperfections get reduced with the advance of technology. Depending on the questions to be investigated, it is not always clear if a deviation mathematical model – real experiment has to be classified as idealization or imperfection. We give a few examples for both, in the framework of the questions to be investigated in the following text.

Idealizations are for instance:

the fluid is inviscid (eliminated);

the cylinders are infinitely long, i.e. end effects can be neglected;

the fluid is homogeneous, i.e. the influence of the aluminum flakes in the oil can be neglected; the influence of gravity can be neglected;

the flow is isothermal, i.e. internal heating of the fluid because of dissipation can be neglected; viscosity and density are constant.

Imperfections are for instance:

the cylinders are not perfectly concentric;

a cylinder was not held rigidly enough and could perform small lateral movements;

the cooling device works reliably only up to certain Reynolds numbers.

To keep the effects both of idealizations and of imperfections reliably negligibly small sometimes required quite some efforts and inventiveness of the experimenters and is discussed in their papers. We shall mention further examples where they played an important role.

Figure 1.1 shows Taylor's apparatus as given in [Tay]. For technical details of later experiments see for instance [Co65, DS, BuKo, AC, ACDH, ALS, DL87, AH92], and the review article [Do92].

Following Taylor, comparison between experiment and mathematical model was always considered important. In some cases, the agreement between computed and measured values of flow parameters was excellent:

comparison of neutral curves of Couette flow by Taylor [Tay], see Fig. 1.4;

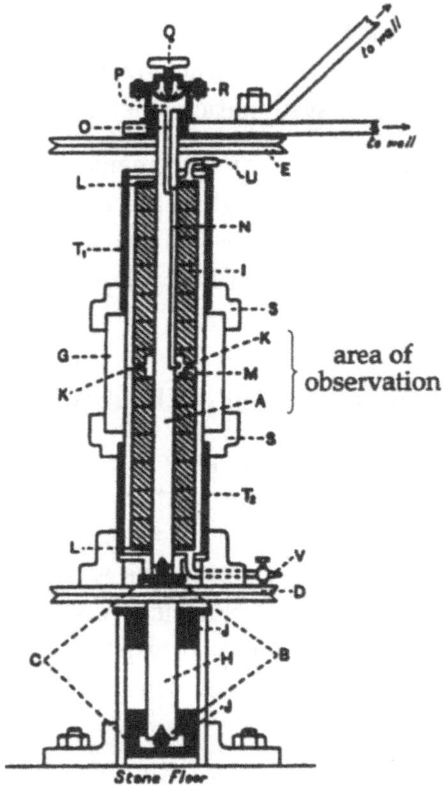

Figure 1.1: Taylor's apparatus [Tay, Fig. 8]. Observations were made in the range of the glass cylinder G.

comparison of torques for Re $\leq$ 2.2 Re$_{cr}$ by Meyer-Spasche/Keller, see Figs. 1.2 and 1.3;
comparison of stability limits for Re $\leq$ 2 Re$_{cr}$ by Riecke/Paap [RP86, Figs.2 and 3], see Fig. 3.37.

In several cases when agreement was poor, convincing explanations were found, and errors could then be reduced considerably. Examples:

- During the investigations of the Eckhaus instability leading to Fig. 3.37, it was found that gravity affected the results. After this was detected, its influence was excluded by using horizontal cylinders instead of vertically standing ones (see the section on the Eckhaus instability).

- In a case when computer simulation and experiment gave qualitatively the same, but quantitatively strongly differing results (the curves obtained were shifted from each other by a fixed amount), it was found that this was due to a systematic error in measuring [ClM85].

The only pronounced modeling disagreement in the Taylor vortex community is whether one can ever neglect the role of the aspect ratio of the experiment or if it always has to be taken into account [BM] (see subsection 'end effects' in this chapter and section 'validity of the numerical model' in the next chapter).

1.1.2 Mathematical description of the experiment

We assume that the experiments are performed with incompressible Newtonian fluids at constant temperature, with constant viscosity and constant density. We thus can describe them by the incompressible **Navier-Stokes equations** with no-slip boundary conditions at the two cylinders. The geometry of the apparatus suggests cylindrical coordinates. Let (r, θ, z) be the spatial coordinates, $\mathbf{v} = (u, v, w)$ the corresponding velocity components, p the pressure, ν the viscosity and ρ the density. The equations then read

$$\frac{du}{dt} - \frac{v^2}{r} = -\frac{\partial}{\partial r}\frac{p}{\rho} + \nu\left(\Delta u - \frac{u}{r^2} - \frac{2}{r^2}\frac{\partial v}{\partial \theta}\right) \tag{1.3}$$

$$\frac{dv}{dt} + \frac{uv}{r} = -\frac{1}{r}\frac{\partial}{\partial \theta}\frac{p}{\rho} + \nu\left(\Delta v - \frac{v}{r^2} + \frac{2}{r^2}\frac{\partial u}{\partial \theta}\right) \tag{1.4}$$

$$\frac{dw}{dt} = -\frac{\partial}{\partial z}\frac{p}{\rho} + \nu\Delta w \tag{1.5}$$

$$0 = \frac{\partial u}{\partial r} + \frac{u}{r} + \frac{1}{r}\frac{\partial v}{\partial \theta} + \frac{\partial w}{\partial z}. \tag{1.6}$$

Here

$$\frac{d}{dt}\cdot = \frac{\partial}{\partial t}\cdot + u\frac{\partial}{\partial r}\cdot + \frac{v}{r}\frac{\partial}{\partial \theta}\cdot + w\frac{\partial}{\partial z}\cdot \tag{1.7}$$

is the material derivative in cylindrical coordinates, and

$$\Delta \cdot = \frac{\partial}{\partial r^2}\cdot + \frac{1}{r}\frac{\partial}{\partial r}\cdot + \frac{1}{r^2}\frac{\partial^2}{\partial \theta^2}\cdot + \frac{\partial^2}{\partial z^2}\cdot. \tag{1.8}$$

There are several ways to *non-dimensionalize* (1.3)–(1.6): we can choose R_1, R_2, $\frac{R_1+R_2}{2}$ or $R_2 - R_1$ as characteristic length $\tilde{L}$, and we can choose $R_1\Omega_1$, $R_2\Omega_2$, $\frac{R_1+R_2}{2}(\Omega_2 - \Omega_1)$, ... as characteristic velocity V. With each of these choices we get a different Reynolds number. Unfortunately, practically all of these choices can be found in the literature, and also several different Taylor numbers. Two very common choices are

$$\tilde{L} = R_1, \quad V = \Omega_1 R_1, \quad \mathrm{Re} = \frac{\Omega_1 R_1^2}{\nu}, \tag{1.9}$$

and

$$\tilde{L} = D := R_2 - R_1, \quad V = \Omega_1 R_1, \quad \mathrm{Re}^\delta = \frac{\Omega_1 R_1 D}{\nu}. \tag{1.10}$$

In their dimensionless form, the equations depend on three parameters:

$$\eta := \frac{R_1}{R_2}, \quad \mu := \frac{\Omega_2}{\Omega_1}, \quad \mathrm{Re}, \tag{1.11}$$

as some of the quantities mentioned earlier are not independent of each other. η, μ, Re are **outer parameters**; their values have to be fixed by the experimenter. Since the behavior of the flows observed strongly depends on the current parameter values, and can be controlled by their changing, they are also called **control parameters**. As we shall see later, there are several **behavior variables** or **inner parameters** in addition: the wavelength λ of the stationary vortices (see for instance Fig. 1.8), the number of circumferential waves of wavy vortices (see Fig. 1.7), etc. The inner parameters are 'chosen by the fluid itself' and differ from steady state to steady state if a system has more than one steady state for one set of control parameters. The appearance of inner parameters is very typical for nonlinear problems, and their number typically changes at bifurcation points. To successfully control an experiment, it is necessary to know its repertoire of possible behaviors, and thus its inner parameters. To better understand the role of the inner parameter λ in the Taylor problem was an important motif for the investigations reported in Chapters 3 and 4.

Equations (1.3)–(1.6) have a stationary solution which does not depend on θ nor on z. It is called **Couette flow**. In dimensional form it is given by

$$\begin{aligned} v(r) &= \mathcal{V}(r) := Ar + \frac{B}{r}, \\ p(r) &= \mathcal{P}(r) := \left(\frac{A^2 r^2}{2} - \frac{B^2}{2r^2} + 2AB \ln r + c\right)\rho, \\ u(r) &\equiv 0 \equiv w(r), \end{aligned} \tag{1.12}$$

with

$$A = -\Omega_1 \eta^2 \frac{1 - \mu\eta^{-2}}{1 - \eta^2}, \qquad B = \Omega_1 \frac{R_1^2 (1 - \mu)}{1 - \eta^2}. \tag{1.13}$$

The expressions for A and B are computed from the boundary conditions at the two cylinders (no slip: the fluid has the velocities $\Omega_i R_i,\quad i = 1,2$ of the walls). The constant c in $\mathcal{P}(r)$ can be chosen arbitrarily, because the pressure p itself does not occur in the Navier-Stokes equations, only its gradient.

In experiments with large aspect ratio and sufficiently small values of Ω_1/ν and Ω_2/ν, Couette flow is a very good approximation to the actual flow in the middle portion of the apparatus. This can be seen, for instance, from the following results:

In the case $\mu = 0$, $\eta = 0.5$ and $\eta = 0.95$ Donnelly/Simon [DS] measured the torque $G(R_2)$ which is felt by a fixed outer cylinder when the inner cylinder is turned with constant angular velocity Ω_1. As long as $\mathcal{V}(R_1)$ was small, they found the linear relation

$$G(R_2) = \frac{4\pi\nu\rho R_1 R_2^2 L_o \mathcal{V}(R_1)}{R_2^2 - R_1^2}, \tag{1.14}$$

as expected. Here L_o is the length in the z-direction which was used in the measurements ($L_o = 5$ cm). Their measured values versus Re^δ from (1.10) are shown as '+' in Figs. 1.2 and 1.3, together with the values obtained for $G(R_2)$ by evaluation of (1.14). The measured values deviate clearly from a straight line for $\mathrm{Re}^\delta \geq \mathrm{Re}^\delta{}_{cr}(\eta)$: Couette flow turns unstable for a Re^δ-value which we call **critical value** $\mathrm{Re}^\delta{}_{cr}$, and which is known to be a bifurcation point [Kir, Ve66]. These figures also show the torque of bifurcated solutions that were obtained by numerical solution of the Navier-Stokes equations. The numerical methods are explained in Chapter 2, the numerical results in Chapter 3. The role of the wavelength λ in these computations is discussed in 'Computations with fixed period $\lambda \approx 2$'.

In that parameter range where the relation is linear, the deviations *measurements – theoretical formula* indicate the size of the deviation *experiment – mathematical model.* The only other errors present are the errors of the measurements, usually estimated and well controlled by the experimenters.

1.1.3 Narrow gap limit and Rayleigh-Bénard problem

Comparison of Figs. 1.2 and 1.3 shows that the experiment seems to be more sensitive to perturbations and imperfections for $\eta = 0.95,\ \mu = 0$ than for $\eta = 0.5,\ \mu = 0$. An analogous observation was made in the numerical study of the discretization errors for the two cases. This is not surprising: If

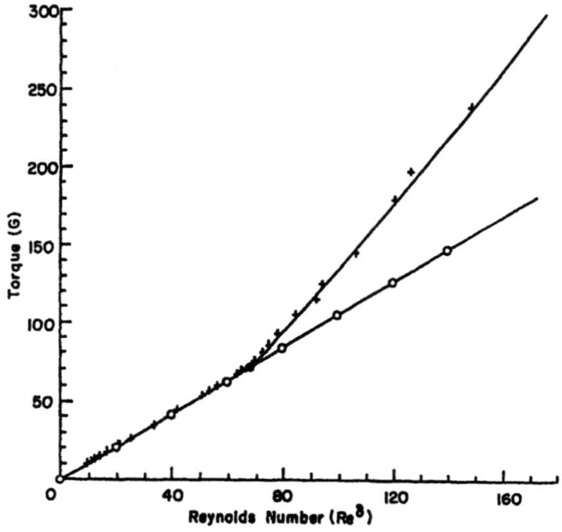

Figure 1.2: Torque (in [dyne cm]) versus Re^δ for $\eta = 0.5,\ \lambda = 1.988,\ \mu = 0$.
+: Measurements of Donnelly/Simon [DS] ($\Gamma = 10$); $-\circ-$: $G(R_2)$ as computed with formula (1.14) for Couette flow; •: computed bifurcation point (M=63); — : computations with the code Tayperio in the nonlinear regime ($M = 31,\ N \leq 4$ for $\mathrm{Re}^\delta \leq 73$; $M = 15,\ N \leq 8$ above) [MK80, Fig.3].

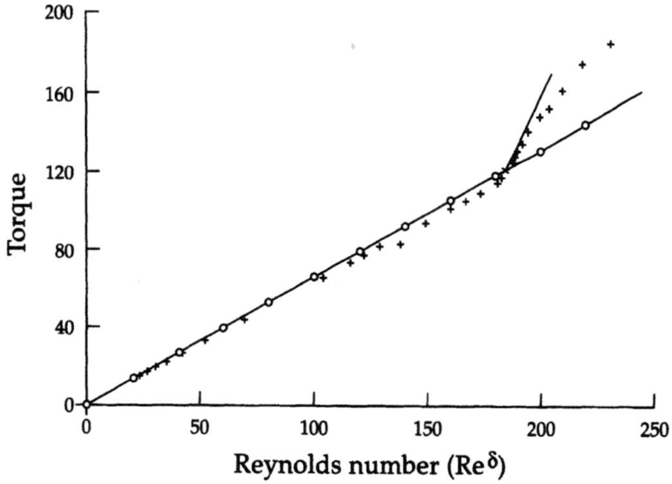

Figure 1.3: Torque versus Re^δ for $\eta = 0.95,\ \lambda = 2.007,\ \mu = 0$.
+: Measurements of Donnelly/Simon [DS] ($\Gamma = 100$); $-\circ-$: $G(R_2)$ as computed with formula (1.14) for Couette flow; — : computations with the code Tayperio in the nonlinear regime. ×: accurate computed bifurcation point (M=80) [MK78, Fig.4.1]. More details in Fig. 2.5 and in Section 3.2.

the two cylinders are arbitrarily close to each other, the only non-singular case is the one with both cylinders turning with the same velocity and in the same direction, i.e. for $\mu = 1$. As we can see from (1.13), $\mathcal{V}$ has a singularity for $\eta \to 1, \quad \mu \neq 1$ fixed. The limit $\eta \to 1$ is well-defined only if performed in the right way, such that all quantities involved stay bounded.

The limiting case $\eta \to 1$ is important for two reasons:

- For $1-\eta \ll 1, \ \mu > 0$ the radial dependence of equations and solutions is weak, and centrifugal effects can be neglected. This allows simplifications of the equations to be solved and thus simplifies application of perturbation methods considerably. In more detail: In the **'narrow gap approximation'**

$$D = R_2 - R_1 \ll \frac{1}{2}(R_2 + R_1) \tag{1.15}$$

is assumed, the term $\frac{1}{r}\frac{\partial \cdot}{\partial r}$ in the differential operator is neglected, and the expression

$$\begin{aligned}
\Omega(r) &= A + B/r^2 \\
&= \Omega_1 \frac{\mu - \eta^2}{1-\eta^2} + \frac{1}{r^2}\Omega_1 \frac{R_1^2(1-\mu)}{1-\eta^2} \\
&= \Omega_1 \left(\frac{1-\eta^2}{1-\eta^2} - \frac{1-\mu}{1-\eta^2} + \frac{R_1^2}{r^2}\frac{1-\mu}{1-\eta^2} \right) \\
&= \Omega_1 \left(1 - (1-\mu) \frac{(r-R_1)(r+R_1)R_2^2}{(R_2-R_1)(R_2+R_1)r^2} \right)
\end{aligned} \tag{1.16}$$

for Couette flow (1.12), (1.13) is simplified to

$$\Omega(r) = \Omega_1 \left(1 - (1-\mu)\frac{r - R_1}{R_2 - R_1} \right) \tag{1.17}$$

[DR, p.82], [Cha, p. 288]. If assumption (1.15) is satisfied, this is a remarkably accurate approximation [Stu, p. 319f]. Taylor used (1.17) in part of his calculations leading to Fig. 1.4, see the text following eq. (1.36).

- If the limit $\eta \to 1, \ \mu \to 1$ is performed *in the right way simultaneously*, the equations for the stationary axisymmetric Taylor problem converge to the 2-dimensional Boussinesq approximation of the Rayleigh-Bénard problem with rigid boundary on top and bottom of the fluid layer and with Prandtl number Pr $= 1$.

It is a mathematical limit, in which the deviation $v - \mathcal{V}$ of the toroidal velocity component is identified with the temperature deviation θ in equation (4.4). But there is also a qualitative correspondence between the two experiments, and this was noticed quite early: Rayleigh already used the physical analogy between the centrifugal instability of rotating fluids and the gravitational instability of stratified fluids when deriving the stability criterion (1.1) [Tay]. It seems that Jeffrey was the first to investigate the relation between the two systems of equations mathematically under the assumption $\eta \to 1$; he reports that Taylor suggested that he do so [Jef28]. Then it was noticed that the governing ordinary differential equations for the critical eigenvalue become identical for $\mu \to 1$ and $\mathrm{Pr} = 1$ [Cha, Stu], [DR, p. 98]. In several of his papers, Busse pointed out that his new results on the Rayleigh-Bénard problem do apply to the Taylor problem (see for instance [BO]). It was shown that the analogy between the two problems carries over to the case when there is an axial flow in the Taylor problem and a horizontal flow in the Rayleigh-Bénard problem, and that the equations for the eigenvalues governing the temporal behavior of the instability in the supercritical regime are closely related [Bue84]. It was shown that the full 2-dimensional equations of the Taylor problem and Rayleigh-Bénard problem correspond to each other, and transformations starting from the narrow gap approximation were given [Bus, Na86]. Nevertheless, it seems that we [LMP] were the first to give an algorithm in which no small terms are neglected, and by which this limit can be performed numerically in a Taylor vortex code for arbitrary gaps by a continuation method. Following [LMP], this transformation is given in Chapter 4.

1.1.4 End effects

Though most of the theory is done on 'infinite cylinders', all experiments have to be done in finite cylinders with real end conditions. As mentioned earlier, many experimenters did a careful analysis of end effects. We give examples:

Burkhalter and Koschmieder [Bu72, BuKo] probably were the first who studied quantitatively the effect of end conditions on Taylor vortex flows. In their experiments with $\mu = 0$, they investigated stationary flows with different boundary conditions on top and bottom:

solid lids, fixed to the rotating inner cylinder,
solid lids, fixed to the non-rotating outer cylinder, and
free surface on top.

Also, they varied the lengths of the cylinders. In all their experiments they measured both the length λ_E of the end cells and the average length λ_f of the inner cells. Thus they could show:

If Re is increased quasistationarily[1], changes in the wavelength λ_f are caused by end effects. An extrapolation to infinitely long cylinders gives constant wavelength $\lambda_f = \lambda_{cr}$ (λ_{cr} is defined in eq. (1.37)). The boundary conditions giving results closest to infinite cylinders were found to be non-rotating lids on top and bottom.

In their experiment with $R_1 = 5.250$ cm, $R_2 = 5.946$ cm, $\eta = 0.883$, Andereck et al. [ALS] varied the aspect ratio Γ (1.2) in the range $20 \leq \Gamma \leq 48$. They mostly used $\Gamma = 30$ in those investigations leading to Fig. 1.5. They report that the value of Γ did not have much influence on the results.

As mentioned earlier, the most advanced concept for end conditions employs ramps [AC]. The experiments by Ahlers, Cannell et al. in Santa Barbara (cited in Fig. 3.37) were ramped. They give the best experimental approximation to infinitely long cylinders known today. The influence of ramps was also investigated theoretically, see for instance [RP87].

Now that the theory for infinitely long cylinders is better understood than at the time when Taylor started his investigations, Benjamin aroused interest in flows in short cylinders with his stimulating experiments [Ben78], and a theory for short finite cylinders is being developed [Ben78, BM, AG78, BH79, Ha80a, Ha80b, Sae, SD80, ND81, Cli, BoKe, Nei84, ClM85]. We call this the **Benjamin problem** in the following.

Though this is an extremely interesting field, we felt that the theory of axisymmetric flows with periodic boundary conditions also needed further development. The concept is simpler and the flows are more perfect, i.e. fewer singularities are unfolded, and especially fewer bifurcations are perturbed. As will be discussed in Chapters 3 and 4, some questions of interest to us are more easily investigated for the Rayleigh-Bénard problem than for the Taylor problem, *because it is more perfect.* We thus confined our own investigations to the more ideal case of axisymmetric flows having periodic boundary conditions and vanishing axial velocity w on top and bottom.

[1] A *variation of parameter* is done *quasistationarily*, if the steps in the control parameter are so small and if they are performed so slowly that the dynamical system attains a sequence of stationary states and never deviates substantially from being stationary.

1.2 Flows between rotating cylinders

In his pioneering investigations, Taylor determined the critical values Ω_i/ν, $i = 1, 2$ of the two cylinders for onset of instability of Couette flow, both experimentally and by calculations. He got very good agreement: the deviations measurements – calculations were less than 5%. One of the marginal curves given by him is repeated here as Fig. 1.4.

Figure 1.5 shows a similar marginal curve for slightly different η in a slightly different normalization: this time, the parameters are the two Reynolds numbers $\mathrm{Re}_i = R_i D\Omega_i/\nu$. In addition, this figure shows the resulting flow if the outer cylinder first attains its final velocity, and then the inner one. If the final state is reached on a different path in parameter space, it is very likely that the flow will attain a different steady state: see the discussion of Figs. 1.6, 1.7, and 1.8 below. Photographs of most of the flows named in Fig. 1.5 are shown in [ALS]; see also [Dy, Figs. 127–131] for Taylor vortices, wavy vortices, spiral turbulent flow and turbulent Taylor vortices.

We now consider the case $\mathrm{Re}_2 = 0$, i.e. $\mathrm{Re}_1 = \mathrm{Re}^\delta$, $\mu = 0$: the outer cylinder at rest. In the parameter planes of Figs. 1.4 and 1.5, this corresponds to only a line. But even under this restriction there is a large variety of different steady states. And we shall vary η now. If Re^δ is increased, first stationary Taylor vortices are generated. Such flows are shown in Fig. 1.8. For which range of Re-values they are stable, strongly depends on the radius ratio η: as is illustrated by Figs. 1.2 and 1.3, stationary flows give way to wavy vortices 'very soon' in the narrow gap case (for $\mathrm{Re}/\mathrm{Re}_{cr} < 1.2$ if $\eta = 0.95$), and they stay axisymmetric for a large range of Reynolds numbers in the wide gap case (up to $\mathrm{Re}/\mathrm{Re}_{cr} > 4$ for $\eta = 0.5$). This is discussed in more detail in the section 'wavy vortices' in Chapter 3.

Experiments by Coles 1965 showed that there are quite a number of different steady states with wavy vortices for identical outer parameters and different inner parameters 'spatial period in the axial direction' and 'temporal period in the angular direction'.

Coles [Co65] was the first to observe that the number of vortices in a given configuration is not uniquely determined by the Reynolds number (η and μ fixed). In his $\eta = 0.874$ apparatus, he could produce up to 28 different wavy vortex equilibria for the same value of Re. Figure 1.7 shows transitions and hysteresis between wavy vortex flows with different numbers of waves. As Coles pointed out, 'the flow pattern was observed to change abruptly, discontinuously, and irreversibly from one state to another at certain well

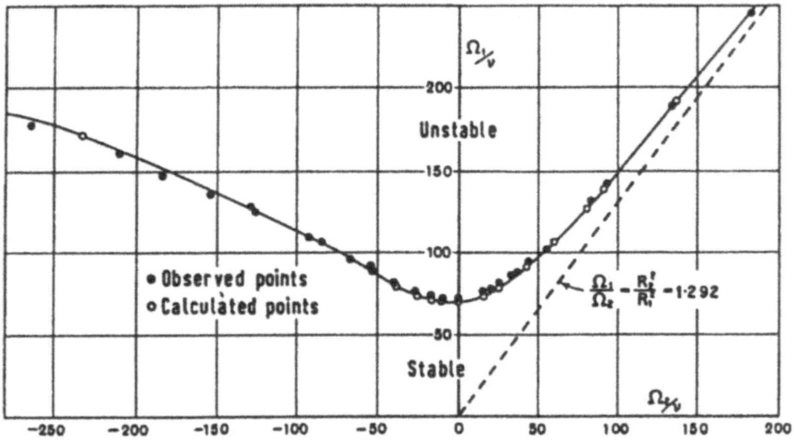

Figure 1.4: Curve of marginal stability for Couette flow, $\eta = 0.8798$, both cylinders rotating. Measurements and calculations by Taylor [Tay, Fig. 18]. The straight line $\mu^{-1} = \eta^{-2} = 1.292$ results from Rayleigh's Circulation Criterion (1.1) for Couette flow.

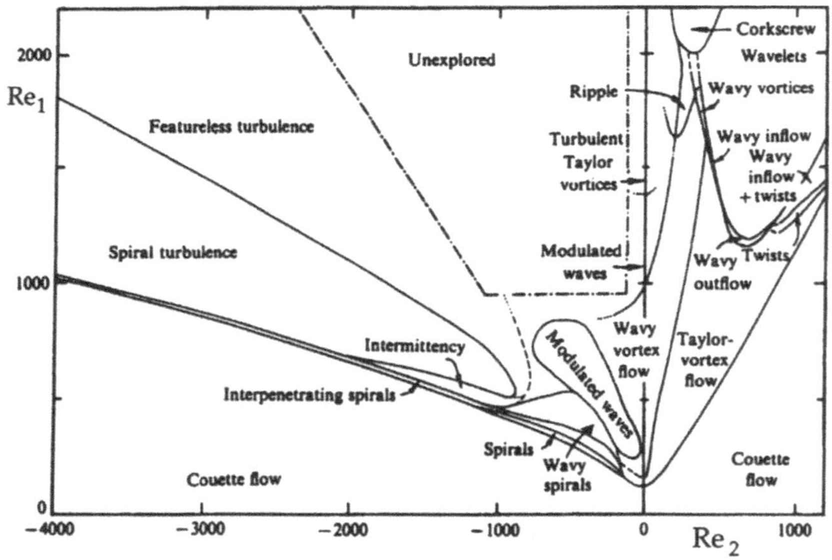

Figure 1.5: Resulting flows for $\eta = 0.883$ if the final Reynolds number $\mathrm{Re}_j = R_j D \Omega_j / \nu$ is obtained first for the outer cylinder ($j = 2$) and then for the inner cylinder ($j = 1$) [ALS, Fig. 1].

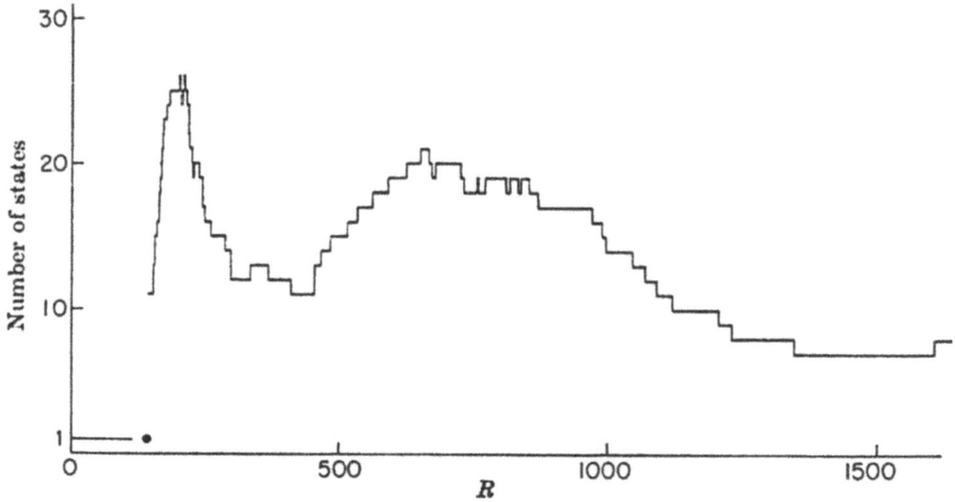

Figure 1.6: Non-uniqueness of steady states: Number of accessible states as a function of the Reynolds number Re^{δ}: $\eta = 0.874$, $\mu = 0$, $\Gamma = 27.9$; $\mathrm{Re}^{\delta} \leq 1500 \approx 14\ \mathrm{Re}^{\delta}{}_{cr}$ [Co65, Fig. 9].

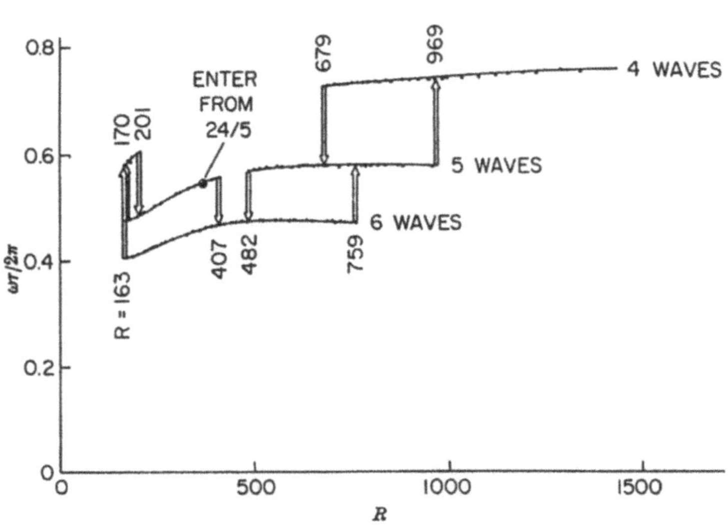

Figure 1.7: Non-uniqueness of steady states: Hysteresis loops in state diagram for 22 Taylor cells [Co65, Fig. 7]: Ratio of period of wave passage τ to period of cylinder rotation $2\pi/\Omega_1$ vs. Re^{δ}. $\eta = 0.874$, $\mu = 0$, $\Gamma = 27.9$; parameter changes were performed quasisteadily.

Figure 1.8: Non-uniqueness of stationary Taylor vortex flows for $\eta = 0.727$, $\mu = 0$, $\mathrm{Re} = \sqrt{9.1}\,\mathrm{Re}_{cr}$, $\Gamma = 51$. The observed wavelengths are $\lambda_f = 1.530$ after sudden start; $\lambda_f = 2.020$ after quasisteady increasing of Re; $\lambda_f = 2.405$ after slow filling; [BuKo, Fig. 8].

defined and repeatable critical speeds'. Figure 1.6 shows the total number of different steady states Coles found. The number strongly varies with the Reynolds number.

Experiments by Snyder 1969 and Burkhalter/Koschmieder 1974 showed that there are also several different stable stationary states for given fixed values of the control parameters. They differ in the inner parameter λ, which we call 'wavelength' or 'period'. As an example we show Fig.8 from [BuKo] as Fig. 1.8. They produced these different flows with experimental tricks: quasistationary change of Re; sudden starts; filling of fluid at constant Re. For more details see the section 'Flows near $\mathrm{Re} = \sqrt{1.5}\,\mathrm{Re}_{cr}$' in Chapter 3.

In the following years, the question of non-uniqueness was mostly studied for axisymmetric stationary Taylor vortex flows, both in experiments [Sny, Bu72, BuKo, Ben78] and in theory [Mey67, DPE, An77]. Chapters 3 and 4 of this text focus on later investigations of existence and variety of stationary flows. How much these investigations increased the knowledge of the flows' repertoire of possible behaviors (i.e. the knowledge of the dependence of the behavior parameter λ on the control parameters) can be seen from a comparison of Fig. 1.8 from 1974 and Fig. 3.37 from 1986: In 1974, flows with different periods were generated, but it was unknown if the full range of λ-values inside the marginal curve of Couette flow (see Fig. 1.9) was possible in principle, i.e. the Eckhaus limit was not known. In 1986, the theoretically accessible range of periods was known from numerical stability investigations of Taylor vortex flows , and the theoretical limit could be verified in the experiment (by using the experimental tricks introduced in 1974: quasistationary change of Re; sudden starts; filling and pouring of fluid at constant Re).

1.3 Stability of Couette flow

As we have seen, Rayleigh used heuristic reasoning to develop his stability criterion (1.1) for rotating inviscid flows. Orr (1907) and Taylor (1923) pioneered the calculation of critical parameter values for onset of instability. Taylor carefully analysed Orr's investigations for *plane* Couette flow, but also those of Sommerfeld, Mises, Hopf and Harrison. He understood that the stability of flows between rotating cylinders is more suitable for obtaining agreement between theoretical and experimental results, because the boundary conditions used in the calculations can be approximated more easily in the experiment. Taylor designed his experiment according to the needs of mathematical analysis.

In his investigations, Taylor already used the 'method of normal modes' which is still very popular in theoretical and computational fluid dynamics, and which we also used in the code Tayperio to be described in the next chapter. Taylor also introduced several other concepts and ideas which were used by many others later on. In the following years, existing methods were further developed and new methods were introduced and applied to various problems in fluid dynamics. There was always a close connection between stability investigations for flows between rotating cylinders and for convecting flows.

In this section we describe a few investigations on the stability of Couette flow. We introduce the principle of linearised stability and the principle of exchange of stability. The main purpose of this section is to provide results to be used later. Stability of Taylor vortex flows is discussed in Chapter 3.

Following Taylor, we first consider the linearised stability of Couette flow with respect to axisymmetric stationary perturbations [Tay], [DR, p. 90ff], [Cha, p. 294ff]. Both the assumption of the validity of the principle of linearised stability and the restriction to stationary axisymmetric perturbations (instead of considering 3-dimensional, time-dependent ones) simplified the investigations substantially and thus made them feasible.

This approach was motivated by evidence from Taylor's experimental observations. Its validity was first justified by its success and later on by mathematical confirmation. The confirmation of its validity [Jo82] also confirmed its expected limitations and provided the corresponding parameter values: the restriction to stationary axisymmetric perturbations is not adequate for certain μ, η-values, $\mu < 0$ (see Fig. 1.12 and the accompanying text). The validity of the principle of linearised stability has recently been questioned by S. Grossmann et al. [GG93, GG94] and Trefethen et al. [TRD93] because the Frechet derivative at stationary states of eqs. (1.3)–(1.6) is not self-adjoint. Thus bifurcations can be subcritical, and perturbations can first grow before they finally decay.

Whenever bifurcations are subcritical or when other solution branches are close without bifurcating, the validity of the principle of linearised stability is likely to be limited to a small basin of attraction of the primary solution[2]. As we shall see in Chapter 3, subcritical bifurcations from Couette flow to Taylor vortex flow occur for $\mu = 0$ only in parameter regimes where Couette flow is unstable anyway. Thus this limitation is not of importance to our investigations. It could be possible, however, that the deviations between experimental and theoretical torque values observed in the subcritical range of Fig. 1.3 were caused by perturbations due to noise. If perturbations grow before they finally decay and if the next perturbation occurs before the previous one has decayed, a theoretically asymptotically stable stationary state can be practically unaccessible in a real system. Also, it could happen that an initially growing perturbation enters the domain of attraction of another stationary state before it is due to decay.

Given an evolution equation

$$\dot{u} = G(u, \lambda) \tag{1.18}$$

[2]It was this effect that complicated the stability investigations for plane Couette flow so much.

in some Banach space $\mathcal{B}$, λ some parameter vector, with stationary state $(\bar{u}, \bar{\lambda})$ satisfying $G(\bar{u}, \bar{\lambda}) = 0$. We say that the **Principle of Linearised Stability** is valid, if the stationary state $(\bar{u}, \bar{\lambda})$ of eq. (1.18) has the same stability as the stationary state $\bar{v} = 0$ of the linearised problem

$$\dot{v} = G_u(\bar{u}, \bar{\lambda})v, \tag{1.19}$$

$G_u(\bar{u}, \bar{\lambda})$ being the Fréchet derivative of G at $(\bar{u}, \bar{\lambda})$.
In the case of ODEs, the principle of linearised stability is valid whenever the 'Theorem of Hartman-Grobman' [GH] is applicable. In the case of the Navier-Stokes equations and Couette flow, its formal validity was shown for the case when all eigenvalues of the linear problem have non-zero real parts [Sa83].

1.3.1 Equations of motion for axisymmetric perturbations

Following Taylor, we linearise the dimensional equations at Couette flow, separate the axisymmetric perturbations with respect to r, t and z, Fourier decompose with respect to t and z ('method of normal modes'), and thus get a system of ODEs for the Fourier coefficients as functions of r. This system is further transformed. When non-dimensionalized, a Taylor number T arises instead of a Reynolds number.
We use the dimensional eqs. (1.3)–(1.6) in cylindrical coordinates (r, θ, z):

$$\begin{aligned}
\frac{du}{dt} - \frac{v^2}{r} &= -\frac{\partial}{\partial r}\frac{p}{\rho} + \nu\left(\Delta u - \frac{u}{r^2}\right), \\
\frac{dv}{dt} + \frac{uv}{r} &= \nu\left(\Delta v - \frac{v}{r^2}\right), \\
\frac{dw}{dt} &= -\frac{\partial}{\partial z}\frac{p}{\rho} + \nu\Delta w, \\
\frac{\partial u}{\partial r} + \frac{u}{r} + \frac{\partial w}{\partial z} &= 0,
\end{aligned} \tag{1.20}$$

with

$$\begin{aligned}
\frac{d\cdot}{dt} &= \frac{\partial\cdot}{\partial t} + u\frac{\partial\cdot}{\partial r} + w\frac{\partial\cdot}{\partial z}, \\
\Delta\cdot &= \frac{\partial^2\cdot}{\partial r^2} + \frac{1}{r}\frac{\partial\cdot}{\partial r} + \frac{\partial^2\cdot}{\partial z^2},
\end{aligned}$$

and

$$u = u_r, \quad v = \mathcal{V}(r) + u_\theta, \quad w = u_z, \quad \frac{p}{\rho} = \frac{\mathcal{P}(r)}{\rho} + \varpi. \tag{1.21}$$

Here $\vec{U}(r) = (0, \mathcal{V}(r), 0, \frac{\mathcal{P}(r)}{\rho})^t$ is the Couette flow (1.12), (1.13) and $\vec{y}(r,z) = (u_r, u_\theta, u_z, \varpi)^t$ is an axisymmetric perturbation.

We use (1.21) in (1.20) and obtain for the first equation:

$$\frac{\partial u_r}{\partial t} + u_r \frac{\partial u_r}{\partial r} + u_z \frac{\partial u_r}{\partial z} - \frac{(\mathcal{V}(r) + u_\theta)^2}{r} = -\frac{\partial}{\partial r}\left(\frac{\mathcal{P}(r)}{\rho} + \varpi\right) + \nu\left(\Delta u_r - \frac{u_r}{r^2}\right). \tag{1.22}$$

We now assume that the perturbation is small and neglect all terms which are nonlinear in the perturbation. Thus we obtain for the first equation

$$\frac{\partial u_r}{\partial t} - \frac{\mathcal{V}(r)^2}{r} - 2\,\frac{\mathcal{V}(r) u_\theta}{r} = -\frac{\partial}{\partial r}\frac{\mathcal{P}(r)}{\rho} - \frac{\partial \varpi}{\partial r} + \nu\left(\Delta u_r - \frac{u_r}{r^2}\right). \tag{1.23}$$

Couette flow satisfies

$$\frac{\mathcal{V}(r)^2}{r} = \frac{\partial}{\partial r}\frac{\mathcal{P}(r)}{\rho}, \tag{1.24}$$

and eq. (1.23) thus simplifies further. Treating the other equations in the same way we obtain:

$$\begin{aligned} \frac{\partial u_r}{\partial t} - 2\,\frac{\mathcal{V}(r) u_\theta}{r} &= -\frac{\partial \varpi}{\partial r} + \nu\left(\Delta u_r - \frac{u_r}{r^2}\right) \\ \frac{\partial u_\theta}{\partial t} + \left(\frac{d\mathcal{V}}{dr} + \frac{\mathcal{V}}{r}\right) u_r &= \nu\left(\Delta u_\theta - \frac{u_\theta}{r^2}\right) \\ \frac{\partial u_z}{\partial t} &= -\frac{\partial \varpi}{\partial z} + \nu \Delta u_z \\ \frac{\partial u_r}{\partial r} + \frac{u_r}{r} + \frac{\partial u_z}{\partial z} &= 0 \end{aligned} \tag{1.25}$$

with the boundary conditions

$$\begin{aligned} u_r(R_i, z) = u_\theta(R_i, z) = u_z(R_i, z) = 0, \quad i = 1, 2, \quad -z_o \le z \le z_o \\ u_*(r, z_o) = u_*(r, -z_o), \quad * = r, \theta, z, \quad R_1 \le r \le R_2. \end{aligned} \tag{1.26}$$

Now we have to solve the linear system (1.25) of partial differential equations with the boundary conditions (1.26). To simplify further, we Fourier expand in z and t. We put:

$$\begin{aligned} u_r(t, r, z) &= e^{pt} u(r) \cos kz, \\ u_\theta(t, r, z) &= e^{pt} v(r) \cos kz, \\ u_z(t, r, z) &= e^{pt} w(r) \sin kz, \\ \varpi(t, r, z) &= e^{pt} \varpi(r) \cos kz. \end{aligned} \tag{1.27}$$

with $kz_o = \pi$.

System (1.25) is linear; thus superposition is possible. Thus we need not work with a Fourier series; it is sufficient to treat each single *mode* separately. We are allowed to assume that p in the term e^{pt} is in $\mathbb{C} \setminus \mathbb{R}$: Real part re e^{pt} and imaginary part im e^{pt} both solve the equations separately and can be superposed as needed.

We use (1.27) in (1.25) and obtain for instance for the equation of continuity:

$$\left(\frac{d}{dr} + \frac{1}{r}\right) u e^{pt} \cos kz + k w e^{pt} \cos kz = 0. \tag{1.28}$$

From this we see that we can expect a non-trivial solution only if the choice of a sin for u_z is accompanied by the choice of a cos in u_r. From the orthogonality of the trigonometric functions we obtain now

$$\left(\frac{d}{dr} + \frac{1}{r}\right) u + kw = 0, \tag{1.29}$$

and this is an ODE involving the Fourier coefficients $u(r)$ and $w(r)$. Using (1.27) on Δu_r we obtain in the same way the expression

$$\begin{aligned}
\underbrace{\left(\frac{d}{dr} + \frac{1}{r}\right)}_{=:\, D_*} \underbrace{\frac{d}{dr}}_{=:\, D} u - k^2 u &= (D_* D - k^2) u \qquad (1.30)\\
&= (D D_* + \frac{1}{r^2} - k^2) u,
\end{aligned}$$

i.e.

$$\Delta u - \frac{u}{r^2} = (D D_* - k^2) u, \tag{1.31}$$

and we thus get the new system:

$$\begin{aligned}
\nu(DD_* - k^2 - \frac{p}{\nu})u + 2\,\frac{V}{r} v &= \frac{d\varpi}{dr}, \\
\nu(DD_* - k^2 - \frac{p}{\nu})v - (D_*\, V)u &= 0, \\
\nu(D_* D - k^2 - \frac{p}{\nu})w &= -k\varpi, \\
D_* u + kw &= 0.
\end{aligned} \tag{1.32}$$

The ansatz (1.27) thus replaces the PDE system (1.25) by the ODE system (1.32). We eliminate w from the last of these equations and obtain for the third one:

$$\frac{\nu}{k^2}(D_* D - k^2 - \frac{p}{\nu}) D_* u = \varpi. \tag{1.33}$$

Using this in the first equation of system (1.32) we get an equation of 4th order, and by further changes this turns into

$$\begin{aligned} \frac{\nu}{k^2}(DD_* - k^2 - \frac{p}{\nu})(DD_* - k^2)u &= 2\,\frac{\mathcal{V}}{r}v, \\ \nu(DD_* - k^2 - \frac{p}{\nu})v &= (D_*\,\mathcal{V})u. \end{aligned} \tag{1.34}$$

Now we make use of $\mathcal{V}(r) = Ar + B/r$ and non-dimensionalize with length scale R_2, viscous time scale R_2^2/ν and the velocity scale ν/R_2 following from that. Replacing in addition u by $(2AR_2^2/\nu)u$, we thus obtain

$$\begin{aligned} (DD_* - a^2 - \sigma)(DD_* - a^2)u &= -Ta^2(\frac{1}{r^2} - \kappa)v, \\ (DD_* - a^2 - \sigma)v &= u, \end{aligned} \tag{1.35}$$

with the boundary conditions

$$\begin{aligned} u(\eta) = u(1) = 0, &\qquad v(\eta) = v(1) = 0, \\ Du(\eta) = Du(1) = 0 &\qquad \text{(because of } w(\eta) = w(1) = 0\text{)}. \end{aligned}$$

Here the following definitions were used:

$$\begin{aligned} a &:= kR_2, \\ T &:= -\frac{4ABR_2{}^2}{\nu^2} = \frac{4\Omega_1{}^2R_1{}^4}{\nu^2}\frac{(1-\mu)(1-\mu/\eta^2)}{(1-\eta^2)^2}, \\ \kappa &:= -\frac{AR_2{}^2}{B} = \frac{1-\mu/\eta^2}{1-\mu}, \\ \sigma &:= pR_2{}^2/\nu. \end{aligned} \tag{1.36}$$

a is called the **wave number**, T is called the **Taylor number**. As there are several different Reynolds numbers in use, there are also several different Taylor numbers in use. σ now describes the dependence of the system on time, e^{pt} was replaced by $e^{\sigma t}$

re $\sigma < 0 \Rightarrow$ Couette flow is stable,

re $\sigma > 0 \Rightarrow$ Couette flow is unstable,

re $\sigma = 0 \Rightarrow$ marginal case.

For given κ, a and T, σ plays the role of an eigenvalue for system (1.35). The problem to be solved now is: find those values of the parameters for which the marginal case occurs.

Taylor calculated the values shown in Fig. 1.4 in the following way: he discretized equations closely related to (1.35) by Bessel functions satisfying the eigenvalue equation $DD_*f = -k^2 f$ and searched for zeroes of a resulting infinite determinant. To simplify his finding of the marginal case, he assumed that the eigenvalue σ is real in the marginal case, i.e. he assumed the validity of the principle of exchange of stability. He fixed values of μ and η and considered the vanishing of the determinant as a 2-parameter problem for the wave number a and the Taylor number T and searched for minimum T as a function of a. Only when evaluating the determinant he used the narrow gap approximation[3] (1.17), and also the Galerkin method to reduce it to a finite determinant.

We say that the **Principle of Exchange of Stability** is valid if one stationary state loses its stability to another stationary state in a smooth way. Let the Principle of Linearised Stability be valid, and let σ be the eigenvalue deciding the stability, i.e. let the flow be stable for re $\sigma < 0$ and unstable for re $\sigma > 0$. The Principle of Exchange of Stability is said to be valid if in the marginal case re $\sigma =$ im $\sigma = 0$.

If re $\sigma = 0$ but im $\sigma \neq 0$ in the marginal case, the principle is not valid. Since we are dealing with equations with real coefficients, the case im $\sigma \neq 0$ is only possible if there is another eigenvalue equal to $\bar{\sigma}$ in the marginal case. This is the case in Hopf bifurcation points.

1.3.2 Computation of marginal curves

In the case $\eta = 0.95$, $\mu = 0$, $\lambda = 2.007$ the principle of exchange of stability is valid: Couette flow is unstable with respect to axisymmetric perturbations for $\mathrm{Re} \geq \mathrm{Re}_{cr}$ (see Fig. 1.3). This case is described by a solution diagram of pitchfork type, Fig. 2.4.

If we fix $\mu = 0$ and η and vary the period λ of one pair of vortices, there is a critical Reynolds number for each λ. We thus get a curve in the (λ, Re) plane, the **marginal curve (curve of neutral stability)** of Couette flow with respect to one pair of vortices. Since we imposed periodic boundary conditions in z, the marginal curve computed for 2-vortex perturbations is not sufficient for deciding on the stability of Couette flow. The periodic repetitions of the 2-vortex curve must also be taken into account, i.e. the marginal curves computed for 4, 6, 8,... vortices. The first 8 such curves

[3] Other authors later on simplified the operator DD_* to DD when dealing with the narrow gap case.

are shown in Fig. 3.25. The **proper marginal curve** is then the **envelope** of all these curves. When we talk about 'the marginal curve' in the following, we shall always mean the marginal curve w.r.t. two vortices.

In the Rayleigh-Bénard problem there is a simple explicit formula for the marginal curve, eq. (4.106). For the Taylor problem it can only be approximated numerically. As discussed earlier, **Taylor** was the first to calculate and measure such curves for several η-values. One of them is shown in Fig. 1.4. **Synge** (1938) developed a method to show that system (1.35) satisfies re $\sigma < 0$ if $\mu > \eta^2$ [DR], [Cha, p. 296, p. 341]. He thus generalised Rayleigh's result to the case $\nu \neq 0$ and to wide gaps. Unfortunately, his new method could not be used to get results for other μ, η-values. **Chandrasekhar and Elbert** in 1958 were the first who calculated marginal curves in the (λ, Re) plane for the wide gap case (without neglecting the r dependence). Following Taylor, they used expansions in Bessel functions. **Kirchgässner** calculated marginal curves in 1961 by using Green's functions. For $\mu \geq 0$ he also gave a mathematical proof for the existence of the eigenvalue σ: this is a proof of the validity of the principle of exchange of stability. In Fig. 1.9 we show the marginal curves for $\mu = 0$ and two η values in the $(\mathrm{Re}^2, 2\pi/\lambda)$-plane as computed by Kirchgässner [Kir]. Both curves have a pronounced minimum at approximately that λ value for which the wavelength of the 2-vortex solution is twice the gap width. We shall call the values of that minimum

$$(\mathbf{Re_{cr}}, \boldsymbol{\lambda}_{\mathbf{cr}}) \tag{1.37}$$

from now on. Note that Re_{cr} and λ_{cr} are still functions of μ and η.

Re_{cr} was already used in the text of Figs. 1.6 and 1.8, as given for the η and μ values of those figures. We repeat in Fig. 1.10 the curves from [Kir] giving the dependence of Re_{cr}^2 on η and μ for $\lambda = 2\pi/3 \approx \lambda_{cr}$ (minimum of the curves shown in Fig. 1.9).

Using the theory of topological degree for completely continuous operators, **Velte** [Ve66] proved in 1966 that the marginal curves are loci of bifurcation of stationary states for arbitrary η and $\mu = 0$, and that the smallest eigenvalue is real and simple. For technical reasons he had to exclude the intersection points of the marginal curves, i.e. the double points that turned out to be very important in our analysis. As we shall see in Chapter 3, these points give rise to secondary bifurcations.

Kirchgässner and Sorger formulated the eigenvalue problem and solved it [KS68]. They presented a constructive method for the calculation of Tay-

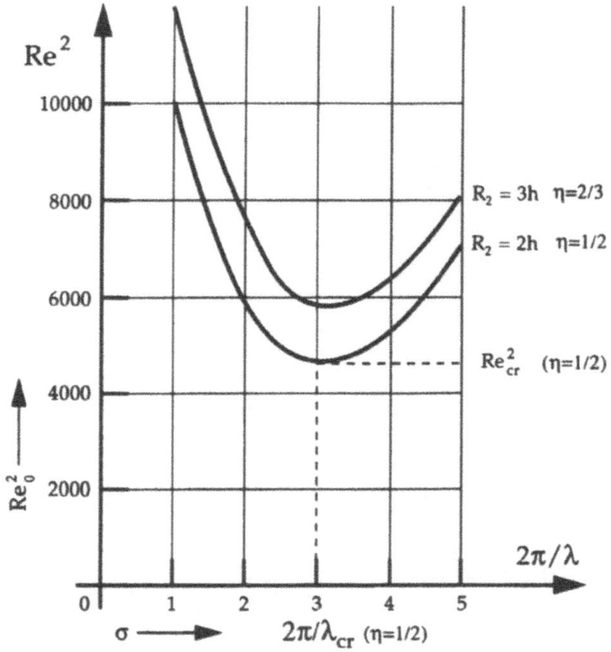

Figure 1.9: Marginal curve in the (Re, λ) plane for two η-values.
Shown are the curves for $\mu = 0$; $\eta = 1/2$ and $\eta = 2/3$ in the $(\mathrm{Re}^2, 2\pi/\lambda)$ plane, [Kir, Fig. 2]. The values of the minimum are called $(\mathrm{Re}_{cr}, \lambda_{cr})$ in the following.

lor vortices as branching solutions of the nonlinear Navier-Stokes boundary value problem [KS69].

Dinar and Keller [DK] gave several views of the marginal surface in the $(\mathrm{Re}^\delta, \lambda, \eta)$ parameter space for $\mu = 0,\ \mathrm{Re}^\delta \leq 250,\ \lambda \leq 12,\ 0 \leq \eta \leq 1$, see Fig. 1.11. They computed them by searching for pitchfork bifurcation points with the code AUTO [Doe] on a system of ordinary differential equations. For $\sigma = 0$, this system is identical with system (1.35), only nondimensionalized in a different way. Usually the Taylor problem is considered only for $0.5 \leq \eta < 1$: as Taylor already observed, the infinite cylinder model is difficult to approximate by experiments with very small η's: 'In that apparatus it was found that if the radius of the inner cylinder was much less than half of that of the outer one, effects due to the ends of the apparatus began to be appreciable and difficult to eliminate, so that the initial motion ceased to be the same as that between two infinite cylinders' [Tay, section 5].

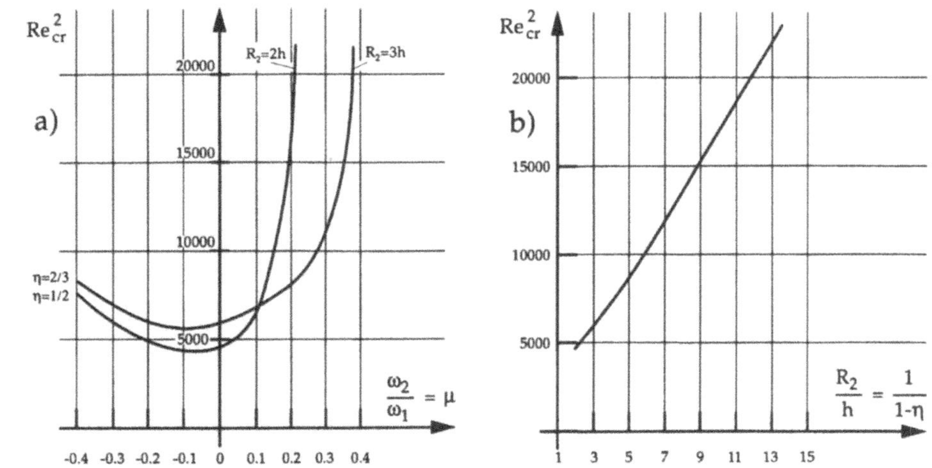

Figure 1.10: a) Re^2_{cr} vs. μ for $\eta = 1/2$ and $\eta = 2/3$; $2\pi/\lambda = 3$ [Kir, Fig. 4]. b) Re^2_{cr} vs. $R_2/D = 1/(1-\eta)$ for $\mu = 0$, $2\pi/\lambda = 3$ [Kir, Fig. 3].

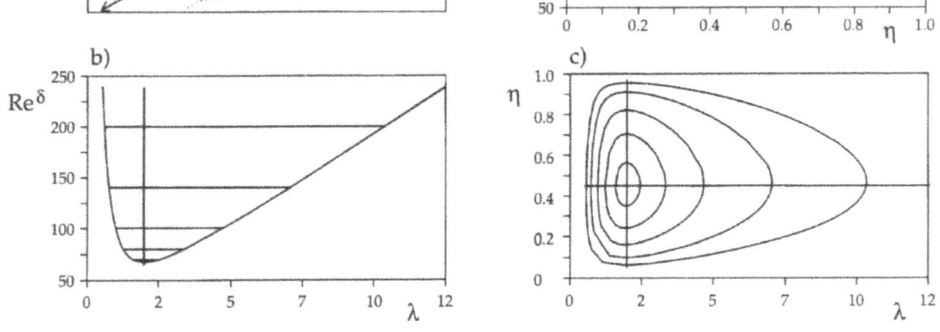

Figure 1.11: Marginal surface in $(\lambda, \eta, \mathrm{Re}^\delta)$ space for $\mu = 0$; [DK, Fig. A1]. a) Qualitative 3D view (minimum at $\mathrm{Re}^\delta = 67.76$, $\lambda = 1.9808$, $\eta = 0.45$). b) Re^δ versus λ section at $\eta = 0.45$. c) Re^δ versus η section at $\lambda = 1.9808$. d) η versus λ sections at $\mathrm{Re}^\delta = 70,\ 80,\ 100,\ 140,\ 200$ (inner to outer).

In our investigations (code Tayperio described in the next chapter and Rayleigh-Bénard problem in Chapter 4) we used the following approach for computing marginal curves: The implicit function theorem states that a nonlinear system of equations is uniquely solvable whenever its Jacobian is non-singular. A singular Jacobian is thus a necessary though not sufficient condition for bifurcation. From earlier work by other authors we knew approximate parameter values for the bifurcation and that this bifurcation is accompanied by an exchange of stability. It was thus sufficient to find the critical parameter values of the discretized systems by evaluating their Jacobian at the discrete analog of the primary flow. In the case when Couette flow is the primary flow, we used bisection or regula falsi to iterate for the critical parameter values for which the Jacobian is singular, i.e. for which the matrix has an eigenvalue $\sigma = 0$. Since we solve the nonlinear equations derived from eqs. (1.3)–(1.6) numerically by a Newton-Gauss continuation method, the Jacobian of the system has to be computed for every parameter value and is thus available. So computation of the marginal curve is no effort and nearly a byproduct. Figures 2.6, 3.24 and 3.25 were computed in that way.

1.3.3 Validity of the principle of exchange of stability

Some results on the validity of this principle were mentioned earlier. Here we give a short review. More details on the investigations before 1982 are given in [DR, Jo82].

- Taylor and many of his successors assumed its validity from experimental evidence, at least for $\mu \geq 0$;
- The analysis for inviscid fluids ($\nu = 0$) did not contradict the assumption of its validity for $\mu > 0$, but spread doubts on its validity for $\mu < 0$ [Tay], [DR, §16];
- Kirchgässner [Kir] proved its validity for $\eta = 1/2$, $\eta = 2/3$ and $\mu \geq 0$ in 1961.
- Using perturbation methods, Davis [Da69] proved its validity for $\eta \approx 1$, $0.68 < \mu < 1$ in 1969;
- Yih [Yih] proved its validity for $\mu > 0$ and all η, $0 < \eta < 1$ in 1972;
- DiPrima, Hall [DPH] showed in 1984 by a counter-example that a main theorem in Yih's second paper [Yih, p. 299] is wrong; this does not have consequences for the validity of the principle;

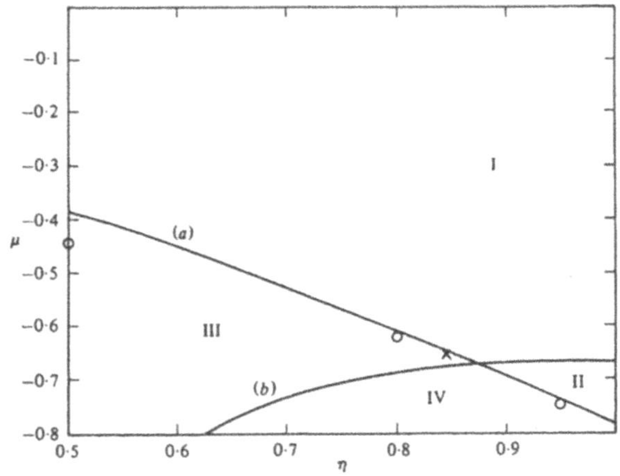

Figure 1.12: Marginal curves for Couette flow, $0.5 \leq \eta \leq 1$, $-0.8 \leq \mu \leq 0$. **Region I**: supercritical Taylor-vortex flow, that is as Re is increased from small values a bifurcation to axisymmetric stable flow is followed by the smooth development of Taylor vortices. In many cases these vortices subsequently become wavy. **Region II**: subcritical Taylor-vortex flow; an axisymmetric transition takes place, but it is followed by a jump to finite-amplitude vortices. Subsequent lowering of Re leads to hysteresis, since stationary vortices are possible for Re-values smaller than critical. **Region III**: non-axisymmetric modes will develop first, so the motion immediately becomes fully 3-dimensional. **Region IV**: the axisymmetric transition is subcritical, but non-axisymmetric modes have a lower critical Re than axisymmetric modes [Jo82, Fig. 1].

- Jones [Jo82] numerically analysed the stability of Couette flow and to which flows Couette flow loses stability in the range $-0.8 \leq \mu \leq 0$, $0.5 \leq \eta < 1$. Result: The principle is valid for $\mu \geq \mu_o(\eta)$, $\mu_o(1) \approx -0.67$, i.e. it is valid in region I of Fig. 1.12. For other (η, μ)-parameter regimes there are subcritical bifurcations to axisymmetric stationary Taylor vortices (see also Chapter 3) and/or bifurcations to non-axisymmetric 3-dimensional flows. More details can be seen from Fig. 1.12. Which flows actually develop in region IV did not follow from Jones' analysis.
- The flows in regions III and IV of Fig. 1.12 were investigated by Chossat and Iooss, Langford et al. and Scheurle [CI94, LTK88, S90, and references therein] (see also Fig. 1.5 for information from experiment to which flows Couette flow loses stability for $\eta = 0.883$ when μ is first brought to its final value and then Re).

Chapter 2

Details of a Numerical Method

In this chapter we describe in detail the code Tayperio. We use it as an example to discuss questions related to the numerical modeling of experiments. When dealing with numerical results on Taylor vortex flows in the next chapter, we shall also include results obtained by other authors with other codes.

In the first section we discuss the numerical model. In the three following sections we look at computational details of the approach used. Finally we discuss the numerical accuracy of the code, focusing on nonlinear effects causing parameter dependence of the numerical error.

2.1 Introduction

In this section we describe the numerical model, review the methods used, discuss the validity of the model, and comment on stability investigations.

2.1.1 Numerical model

We intend to study stationary axisymmetric flows of a viscous, incompressible fluid between long, concentric cylinders. The inner cylinder with radius R_1 rotates at the speed $\Omega_1 R_1$, and the outer cylinder with radius R_2 at the speed $\Omega_2 R_2$. End effects are neglected, i.e. the cylinders have length H in the axial direction, $H \gg R_2 - R_1$, and we assume that there are solutions satisfying periodic boundary conditions in the axial direction. A sketch of the computational domain is shown in Fig. 2.1.

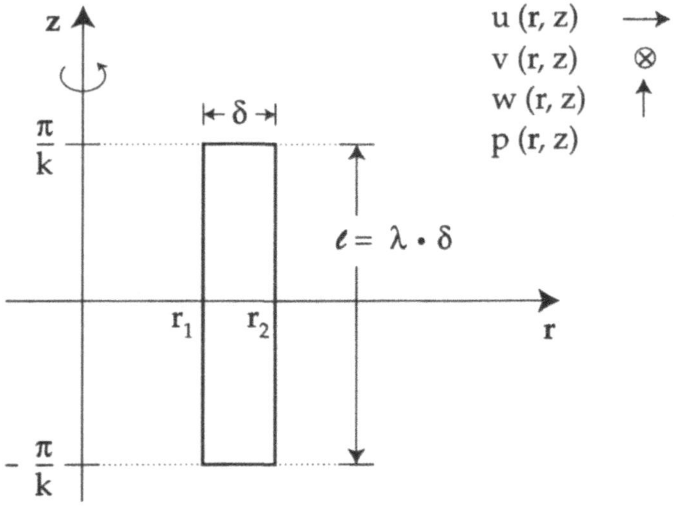

Figure 2.1: Sketch of computational domain. u, v, w are the radial, azimuthal and axial velocity components, resp., p is the reduced pressure.

With these assumptions, the Navier-Stokes equations used to compute such flows read in dimensionless form

$$\begin{aligned} u_r + \frac{1}{r}u + w_z &= 0, \\ \Delta u - \frac{1}{r^2}u - p_r &= \text{Re}\,(uu_r + wu_z - \frac{v^2}{r}), \\ \Delta v - \frac{1}{r^2}v &= \text{Re}\,(uv_r + wv_z + \frac{uv}{r}), \\ \Delta w - p_z &= \text{Re}\,(uw_r + ww_z). \end{aligned} \tag{2.1}$$

Here lengths have been scaled by R_1 and velocities by $\Omega_1 R_1$, as in (1.9). The Reynolds number is given by

$$\text{Re} := \Omega_1 R_1^2/\nu \,. \tag{2.2}$$

ν is the kinematic viscosity of the fluid, and Δ is the Laplacian,

$$\Delta := \frac{\partial^2}{\partial r^2} + \frac{1}{r}\frac{\partial}{\partial r} + \frac{\partial^2}{\partial z^2} = \frac{1}{r}\frac{\partial}{\partial r}(r\frac{\partial}{\partial r}) + \frac{\partial^2}{\partial z^2} \,.$$

There are four independent parameters in the problem: the Reynolds number Re, the rotation rate $\mu = \Omega_2/\Omega_1$, the radius ratio $\eta = R_1/R_2$, and the axial period. The inner parameter 'axial wavelength' of the experiment enters the numerical model as integer fraction of the control parameter 'period'. This will be discussed further in the next chapter. The dimensionless axial period normalized by the inner radius R_1 according to (1.9) is called ℓ, the period based on gap width according to (1.10) is called λ. The dimensionless gap width $\delta := (R_2 - R_1)/R_1$ is related to η by $\delta = \eta^{-1} - 1$ and the wave number k is related to ℓ by $\ell = 2\pi/k$ and to λ by $\lambda = 2\pi/(k\delta)$, see also (2.75).

With these definitions, the boundary conditions can be formulated as

$$\begin{array}{rcll} u(1,z) & = & u(1+\delta,z) = 0\,, & \\ w(1,z) & = & w(1+\delta,z) = 0\,, & -\frac{\pi}{k} \le z \le \frac{\pi}{k}\,, \\ v(1,z) = 1, & & v(1+\delta,z) = \mu/\eta\,, & \\ u(r,-\frac{\pi}{k}) & = & u(r,\frac{\pi}{k})\,, & \\ v(r,-\frac{\pi}{k}) & = & v(r,\frac{\pi}{k})\,, & 1 \le r \le 1+\delta\,. \\ w(r,-\frac{\pi}{k}) & = & w(r,\frac{\pi}{k}) = 0\,, & \end{array} \tag{2.3}$$

These boundary conditions imply that the velocity of the fluid at the walls of the two cylinders is the same as the velocity of the cylinders, i.e. they are no-slip on the two cylinders. They enforce

$$\mathbf{v} \cdot \mathbf{n} = 0 \tag{2.4}$$

on all boundaries, i.e. no fluid can leave the domain of computation on the top and bottom. Theory tells us that the surface integral of $\mathbf{v} \cdot \mathbf{n}$ has to vanish, because of $\operatorname{div} \mathbf{v} = 0$ and the divergence theorem. In practice the stronger local condition (2.4) conveniently excludes the continuum of solutions we would get by shifting one given solution by some arbitrary amount in the z-direction. Whether it also excludes solutions that are more interesting will be discussed in the subsection on 'validity'.

Since system (2.1) contains only the gradient of the pressure, not the pressure itself, we also need a condition fixing the constant up to which p is determined. We chose

$$\int_{-\frac{\pi}{k}}^{\frac{\pi}{k}} p(r_o, z)\, dz = 0, \tag{2.5}$$

r_o to be given in (2.30). Equations (2.1), (2.3) and (2.5) form a system of equations which is locally uniquely solvable for almost all parameter values.

2.1.2 Numerical methods

We now give a short review of the numerical methods used. More details are given in the following sections.

Discretization: Because of the periodic boundary conditions it seems reasonable to use a Fourier decomposition in the axial direction, together with a Galerkin method. In the radial direction we use equidistant central finite differences of second order. Thus system (2.1)–(2.5) is approximated by a nonlinear algebraic system

$$\mathbf{F}(\mathbf{x}, \mathrm{Re}, k, \eta, \mu) = A(k, \eta)\mathbf{x} - \mathrm{Re}\ \mathbf{f}(\mathbf{x}; k, \eta) + \mathbf{b}(\eta, \mu) = 0. \tag{2.6}$$

Here the vector $\mathbf{x}$ denotes the unknowns, i.e. the values of the Fourier coefficients at the radial grid points, while k, Re, η and μ are the parameters introduced earlier.

Most of the computations were performed with 33 radial grid points and $N \leq 28$ Fourier components. Details are given in the captions of the figures. The **accuracy** of this approximation is second order with respect to the finite differences. Its convergence with respect to the Fourier decomposition is exponential, if the approximated solutions are smooth enough.

To get a good estimate of the actual size of the discretization errors (i.e. of the actual values of the bounding constants), we repeated many computations with several different discretizations. We found that the actual size of the errors strongly depends on the parameter values. And for the same values of the control parameters Re, η, μ they usually are different on different solution branches. We also observed a (weak) nonlinear interaction of the discretization errors of the two methods. Not much seems to be known about that interaction theoretically. We shall discuss the discretization errors and our numerical investigations of their sizes in detail in the section on 'numerical accuracy'. In particular we shall explain there why the Fourier discretization error is strongly influenced by the bifurcation pattern of the solution set.

Continuation method: We investigate the solution set of (2.6) by the method of continuation as developed by H.B. Keller [Ke77, Ke87], both with respect to the Reynolds number Re and with respect to the wave number k. In this method, the pseudo-arclength s is used to parameterize the solution curve

$$\mathbf{y}^t(s) = (\mathbf{x}^t, \mathrm{Re})[s] \quad \text{or} \quad \mathbf{y}^t(s) = (\mathbf{x}^t, k)[s]\,. \tag{2.7}$$

In pseudo-arclength continuation with respect to a parameter β ($\beta = \text{Re}$ or $\beta = k$), we add one more unknown and equation, that is: we parameterize the curve $(\mathbf{x}^t, \beta)$ by pseudo-arclength s. Once we know one solution $\mathbf{y}^t(s_o) = (\mathbf{x}^t, \beta)[s_o]$ and the tangent vector $(\dot{\mathbf{x}}^t, \dot{\beta})[s_o]$, we take $\mathbf{y_o}^t = (\mathbf{x}^t, \beta)[s_o] + (s - s_o)(\dot{\mathbf{x}}^t, \dot{\beta})[s_o]$ as start vector for computing a 'neighboring solution' $\mathbf{y}^t(s_1) = (\mathbf{x}^t, \beta)[s_1]$ by the (simplified) **Newton-Gauss method**. This allows us to compute a whole connected component of the solution set over the (Re, k) parameter plane whenever one solution in that connected component is known: fold points can easily be passed, and at simple bifurcation points it is easy to switch branches.

Evaluations: When the system (2.6) has been solved and $\mathbf{x}$ is known, the Fourier coefficients $p_\nu(r_\mu), u_\nu(r_\mu), v_\nu(r_\mu), w_\nu(r_\mu)$ are known at all radial grid points r_μ. A Fast Fourier transformation then gives the reduced pressure $p(r, z)$ and the velocity components $u(r, z)$, $v(r, z)$ and $w(r, z)$ at the grid points of the (r, z) computational domain. The stream function $\psi(r, z)$ is also computed. ψ is defined by

$$u = \frac{1}{r}\frac{\partial \psi}{\partial z}\,, \qquad w = -\frac{1}{r}\frac{\partial \psi}{\partial r}\,. \tag{2.8}$$

These functions are then evaluated, i.e. their level lines are plotted for fixed parameter values, their kinetic energies and their torque are computed, and several quantities are plotted as functions of the parameters varied.

In its final form the code Tayperio consists of 9 main programs and about 40 subroutines. It has about 5000 lines, not counting plotting software. Computations were performed on several different machines, like IBM 370/158, Cyber 175, and several different Crays.

2.1.3 Validity of the model

Equations (2.1)–(2.3) allow solutions which are mirror-symmetric with respect to the plane $z \equiv 0$. Our Fourier decomposition is such that all computed solutions exhibit this sort of symmetry. Today this has to be considered a restriction. Solutions with another symmetry have been detected in the Rayleigh-Bénard problem ('mean component solutions', [BO, Bus87]). And a branch of non-symmetric solutions bifurcating from mirror-symmetric solutions has been found in the **Benjamin problem** (finite, short cylinders with rigid boundaries) [BoKe, Fig. 4d]. Since the Taylor problem is closely related to these two problems, we do expect that such unsymmetric solutions also exist in the Taylor problem and have been excluded by our ansatz (2.11). The number and variety of mirror-symmetric solutions which we found for some parameter values is so large that we sometimes considered this restriction to be advantageous: it made it easier to understand the structure of the remaining solution manifold. For $\text{Re} = 3.65\ \text{Re}_{cr}$, $\eta = 0.727$, $\mu = 0$, for instance, we found more than 21 solutions for $\lambda = 2.5$, more than 25 fold points for varying λ, and period doubling and tripling bifurcations.

It is controversial if equations (2.1)–(2.3) are suited at all to model flows in the middle portion of long cylinders. Comparisons of computed values with measurements have shown that they are even *well* suited for describing axisymmetric steady Taylor vortex flows in the middle portion of a long apparatus:
comparison of marginal curves of Couette flow by G.I. Taylor, Fig. 1.4 and [Tay, Fig. 17]; comparison of torques for $\text{Re} \leq 2.2\ \text{Re}_{cr}$ by Meyer-Spasche/Keller, Figs. 1.2 and 1.3; comparison of stability limits for the basic Taylor vortex flow for $\text{Re} \leq 2\text{Re}_{cr}$ by Riecke/Paap, Fig. 3.37 and [RP86, Fig.2]. As discussed in the previous chapter, all experimenters using the infinite cylinder approach made sure that their results do not depend noticeably on the aspect ratio.

For most of our investigations we fixed η at $\eta = 0.727$, a value that allows steady axisymmetric flows up to $\text{Re} \geq 9\text{Re}_{cr}$ for $\lambda = 2$. This was shown both theoretically [Jo81] and experimentally [BuKo].

Does the condition

$$w(r, \frac{\pi}{k}) = 0 \tag{2.9}$$

exclude non-trivial *axisymmetric* solutions from being computed? To date, this is not known, despite the investigations mentioned near eq. (3.1).

The following property of our approach should be noticed: though we sometimes call λ the 'wavelength', we actually use it as the length of the period. That period can contain substantially more than two vortices. In the investigations for $\text{Re} = 2\ \text{Re}_{cr}$, for instance, we computed up to 16 not equally spaced vortices within one period λ. Many of the computed solutions have non-planar boundaries for most of their vortices (see Figs. 3.12, 3.13, 3.28, 3.33, for instance).

Condition (2.9) does allow us in principle to model certain finite cylinder solutions: if solutions of eqs. (2.1)–(2.3) do satisfy $v(r, \pi/k) \equiv 0$ and $u(r, \pi/k) \equiv 0$ *in addition*, they are solutions of the Benjamin problem as well. We never found such solutions. Maybe because we did not search for them systematically, maybe because such solutions would be too peculiar.

From the ansatz (2.11) it follows that these solutions would also satisfy Schaeffer's boundary conditions

$$\partial u/\partial z = \partial v/\partial z = 0, \qquad w = 0. \tag{2.10}$$

In their investigations reported in [BoKe], Bolstad et al. computed homotopies between solutions satisfying the boundary conditions (2.10) and solutions of the Benjamin problem. They never found flows solving both problems simultaneously [Bol].

2.1.4 Stability

For the convergence properties of the Newton iterations with Gauss elimination it does not make any difference if the computed flows are stable or not. This is advantageous when the computation of unstable flows is desirable.

In the Benjamin problem, the Newton-Gauss method provides a very convenient criterion for linear stability: the sign of the determinant of the Jacobian indicates whether the flow is stable or not [Ben76, Cli]. This is so convenient since the algorithm requires the computation of the Jacobian in each Newton step, and the Gauss elimination provides the sign of its determinant.

In the Taylor problem, such a simple stability criterion does not exist. In a narrow gap, stationary Taylor vortex flows lose stability to wavy vortices, see the text near Fig. 3.35. The flow ceases to be stationary and axisymmetric, and that does not cause a singularity of the Jacobian in an axisymmetric numerical model.

In a wide gap, Taylor vortices lose stability owing to the Eckhaus instability or the short-wavelength instability. At the short-wavelength Eckhaus limit, one pair of vortices shrinks and fades away, while the remaining vortices adapt and grow accordingly. At the long wavelength Eckhaus limit, one pair of vortices is created and grows until all vortices have the same, shorter wavelength. The Eckhaus instability is a *bulk instability*, i.e. it sets in globally. An analysis using only part of the domain and periodic boundary conditions cannot detect it by such simple means as a sign change of the determinant: the volume between the cylinders corresponding to this part of the domain changes its size when this instability sets in. A full stability analysis using Floquet theory is necessary and appropriate [Jo81, RP86]. Figure 3.37 shows how well the computations of Riecke/Paap using this approach agree with the measurements of Domingues-Lerma et al.

It is quite possible that some of the solutions discussed in the next chapters are 'absolutely' unstable, i.e. that there are no experimental side conditions which could allow such flows to exist in the material world. The investigation of such flows is nevertheless of mathematical and physical interest: the neutral curves for both the Eckhaus and the short-wavelength instabilities are strongly influenced by the secondary bifurcations. This is illustrated by Fig. 3.40 and will be discussed at the end of the next chapter.

2.2 The discretized system

We now discuss in more detail the numerical methods leading to system (2.6), and their advantages and disadvantages.

2.2.1 Discretization in the axial z-direction

We seek approximate solutions of (2.1) - (2.5) of the form

$$\begin{aligned} u(r,z) &= \sum_{\nu=1}^{N} u_\nu(r)\cos\nu kz \qquad (2.11)\\ v(r,z) &= \sum_{\nu=0}^{N} v_\nu(r)\cos\nu kz \\ w(r,z) &= \sum_{\nu=1}^{N} w_\nu(r)\sin\nu kz \\ p(r,z) &= \sum_{\nu=0}^{N} p_\nu(r)\cos\nu kz \end{aligned}$$

for $r_1 := 1 \le r \le 1+\delta =: r_2$ and $-\pi \le kz \le \pi$.

Using this ansatz means employing the 'method of normal modes'. As shown in the previous chapter, it was already employed by Taylor [Tay] and Chandrasekhar [Cha]. From their investigations it is known that Taylor vortices bifurcating from Couette flow can be approximated quite well by a single mode ($N = 1$). Thus it was reasonable to hope that few modes would be sufficient for computing moderately supercritical flows.

Ansatz (2.11) implies that u, v, and p are symmetric w.r.t. $z = 0$, and that w is skew-symmetric in z.

The assumption that w is skew-symmetric for symmetric u, v, p follows directly from the equation of continuity. The assumption of symmetry for

u, v, and p, however, is a severe restriction. This was already discussed in the subsection on the validity of our model.

Convergence of the Fourier Series: On smooth solutions, Fourier series converge exponentially [GO]:

Let $f \in C^{n-1}[0, 2\pi]$ be a periodic function, $f^{(n)}$ still integrable.

$$g(x) := \sum_{k=-\infty}^{\infty} a_k e^{ikx} \text{ with } a_k := \frac{1}{2\pi} \int_0^{2\pi} f(x) e^{-ikx}\, dx. \tag{2.12}$$

Then

$$|a_k| \ll |1/k^n| \text{ for } k \to \infty. \tag{2.13}$$

If we define

$$g_N(x) := \sum_{k=-N}^{N} a_k e^{ikx}, \tag{2.14}$$

we get

$$f(x) - g_N(x) = \mathcal{O}(1/N^n) \tag{2.15}$$

for fixed x and $N \to \infty$. If $f \in C^\infty[0, 2\pi]$, convergence is **exponential** for x fixed. It should be noticed, however, that the convergence is not uniform in neighborhoods of those x in which f is not continuously differentiable. Computations with too few Fourier modes can suffer from the Gibbs phenomenon when $f'(x)$ gets large, see Fig. 2.8 in the section on numerical accuracy.

Advantages of the Fourier decomposition are:

- The solutions are expanded into a system of orthogonal functions: if we recompute a solution with $N_2 > N_1$ modes, the coefficients computed previously do not change. Thus it is very easy to apply the **multi-grid principle** and perform the computations with the minimum number of modes necessary: whenever it seems necessary to increase the number of modes, N_1 is replaced by N_2 and the coefficients computed with N_1 are supplemented by zeroes. No interpolations or re-computations are necessary (see the section 'Numerical accuracy').
- Many of the unperturbed bifurcation points of the exact equations (2.1)–(2.5) are approximated by unperturbed bifurcation points of the numerical model equations (2.6). This is discussed further in the subsections on 'periodicity' and on 'numerical accuracy'.

- It is easy to evaluate the computed results component wise and thus to get new insights about the bifurcation pattern of the solution manifold (example: comparison of our computations with the sketch by Busse in Fig. 3.18).

We apply the ansatz (2.11) to eqs. (2.1)–(2.5) and thus get a system of ODEs. Example: The equation of continuity reads now

$$\begin{aligned}
&\frac{\partial}{\partial r}\sum_{\nu=0}^{N} u_\nu(r)\cos\nu kz + \frac{1}{r}\sum_{\nu=0}^{N} u_\nu(r)\cos\nu kz + \frac{\partial}{\partial z}\sum_{\nu=0}^{N} w_\nu(r)\sin\nu kz \qquad (2.16)\\
&\quad = \sum_{\nu=0}^{N} u_\nu'(r)\cos\nu kz + \frac{1}{r}\sum_{\nu=0}^{N} u_\nu(r)\cos\nu kz + \sum_{\nu=0}^{N} w_\nu(r)\nu k\cos\nu kz\\
&\quad = \sum_{\nu=0}^{N}\left(u_\nu'(r) + \frac{1}{r}u_\nu(r) + w_\nu(r)\nu k\right)\cos\nu kz = 0
\end{aligned}$$

and reduces thus to a system of $N+1$ ODEs in r for the Fourier coefficients u_ν and w_ν. Here $'$ stands for differentiation w.r.t. r. ◇

The other equations of (2.1)–(2.5) are discretized in a similar way, but they contain products of sums. These have to be transformed using the addition formulas for trigonometric functions and the Galerkin method to neglect those terms containing entries with index ν satisfying $N < \nu \leq 2N$. For instance we get

$$\begin{aligned}
\left(\sum_{\nu=0}^{N} a_\nu\cos\nu z\right)\left(\sum_{\nu=0}^{N} b_\nu\cos\nu z\right) &= a_o b_o + \frac{1}{2}\sum_{j=1}^{N} a_j b_j + \qquad (2.17)\\
&+\frac{1}{2}\sum_{\nu=1}^{N}\left(\sum_{j=0}^{\nu} a_j b_{\nu-j} + \sum_{j=\nu}^{N}(b_j a_{j-\nu} + a_j b_{j-\nu})\right)\cos\nu z.
\end{aligned}$$

The ansatz (2.11) thus approximates eqs. (2.1)–(2.5) by a system of $4N+2$ non-linear 2-point-BVPs

$$p_o' = \text{Re}\ f_1(\mathbf{u},\mathbf{u}',\mathbf{v},\mathbf{w};k;0,r), \qquad (2.18)$$

$$-u_\nu'' - \frac{1}{r}u_\nu' + ((\nu k)^2 + \frac{1}{r^2})u_\nu + p_\nu' = \text{Re}\ f_1(\mathbf{u},\mathbf{u}',\mathbf{v},\mathbf{w};k;\nu,r),$$

$$u_\nu' + \frac{1}{r}u_\nu + \nu k w_\nu = 0, \qquad (2.19)$$

$$-v_o'' - \frac{1}{r}v_o' + \frac{1}{r^2}v_o = \text{Re}\ f_2(\mathbf{u},\mathbf{v},\mathbf{v}',\mathbf{w};k;0,r), \tag{2.20}$$

$$-v_\nu'' - \frac{1}{r}v_\nu' + ((\nu k)^2 + \frac{1}{r^2})v_\nu = \text{Re}\ f_2(\mathbf{u},\mathbf{v},\mathbf{v}',\mathbf{w};k;\nu,r),$$

$$-w_\nu'' - \frac{1}{r}w_\nu' + (\nu k)^2 w_\nu - \nu k p_\nu = \text{Re}\ f_3(\mathbf{u},\mathbf{w},\mathbf{w}';k;\nu,r), \tag{2.21}$$

for $\nu = 1,\dots,N$, with $\mathbf{u} = (u_\nu)_{\nu=1,\dots N}$ etc., and

$$f_1(\mathbf{u},\mathbf{u}',\mathbf{v},\mathbf{w};k;0,r) = \frac{1}{r}v_o^2 - \frac{1}{2}\sum_{j=1}^{N}(u_j u_j' - \frac{1}{r}v_j^2 - jku_jw_j), \tag{2.22}$$

$$f_1(\mathbf{u},\mathbf{u}',\mathbf{v},\mathbf{w};k;\nu,r) =$$

$$\frac{2}{r}v_ov_\nu - \frac{1}{2}\left(\sum_{j=1}^{\nu-1}(u_ju_{\nu-j}' - \frac{1}{r}v_jv_{\nu-j} + (\nu-j)ku_{\nu-j}w_j)\right.$$

$$\left. + \sum_{j=\nu+1}^{N}(u_ju_{j-\nu}' + u_j'u_{j-\nu} - \frac{2}{r}v_jv_{j-\nu} - (j-\nu)ku_{j-\nu}w_j - jku_jw_{j-\nu})\right),$$

$$f_2(\mathbf{u},\mathbf{v},\mathbf{v}',\mathbf{w};k;0,r) = -\frac{1}{2}\sum_{j=1}^{N}(u_j(v_j' + \frac{1}{r}v_j) - jkv_jw_j),$$

$$f_2(\mathbf{u},\mathbf{v},\mathbf{v}',\mathbf{w};k;\nu,r) =$$

$$-u_\nu v_o' - \frac{1}{r}u_\nu v_o - \frac{1}{2}\left(\sum_{j=1}^{\nu-1}(u_jv_{\nu-j}' + \frac{1}{r}u_jv_{\nu-j} + (\nu-j)kv_{\nu-j}w_j)\right.$$

$$+ \sum_{j=\nu+1}^{N}(u_jv_{j-\nu}' + v_j'u_{j-\nu} + \frac{1}{r}(v_ju_{j-\nu} + u_jv_{j-\nu})$$

$$\left. - (j-\nu)kv_{j-\nu}w_j - jkv_jw_{j-\nu})\right),$$

$$f_3(\mathbf{u},\mathbf{w},\mathbf{w}';k;\nu,r) = -\frac{1}{2}\left(\sum_{j=1}^{\nu-1}(u_jw_{\nu-j}' + jkw_{\nu-j}w_j)\right.$$

$$\left. + \sum_{j=\nu+1}^{N}(w_j'u_{j-\nu} - u_jw_{j-\nu}' - \nu kw_{j-\nu}w_j)\right).$$

From (2.19) and the boundary conditions (2.3a) it follows that $u_o(r,z) \equiv 0$. This has been used in the ansatz (2.11a). It reduces the number of equations by M, the number of grid points in r, and makes the pattern of equations and unknowns more regular ($\Rightarrow$ easier to program).

2.2.2 Discretization in the radial r-direction

Equations (2.18) - (2.21) are discretized by central finite differences in the r-direction. Let M be the number of interior grid points, $h := \frac{\delta}{M+1}$ the mesh width. First and second derivatives are approximated by

$$\begin{aligned} u'(r) &\approx \frac{1}{2h}(u(r+h) - u(r-h)), && (2.23)\\ -u''(r) &\approx \frac{1}{h^2}(2u(r) - u(r-h) - u(r+h)). && (2.24) \end{aligned}$$

The accuracy of this approximation is $\mathcal{O}(h^2)$, see textbooks on numerical analysis. This theoretically expected convergence rate was verified for our code ($\Rightarrow$ no programming errors), see the section on numerical accuracy, especially Figs. 2.5 and 2.6.

Formula (2.23) is numerically unstable on initial value problems of first order. We don't use it on an initial value problem, but on first order equations which are generated from the equation of continuity (2.1a). At the boundary we couple them with the difference equations from (2.1b) when formulating the 'boundary conditions' for the pressure. This worked very well and we never had any stability problems.

2.2.3 Boundary conditions

From the boundary conditions (2.3) and condition (2.5) we get the following conditions for the discrete system:

$$\begin{aligned} u_\nu(r_i) &= 0, \\ v_o(r_1) &= 1, \quad v_o(r_2) = \mu/\eta \\ v_\nu(r_i) &= 0, \\ w_\nu(r_i) &= 0, \qquad i = 1,2, \quad \nu = 1,\ldots,N \qquad (2.25)\\ p_o(r_o) &= 0. \end{aligned}$$

Due to the nature of the Navier-Stokes equations, we are not allowed to impose any boundary conditions on the pressure p. By counting the numbers of equations and unknowns, we find that there are more unknowns than equations. Since we want to solve the nonlinear system by linearization (Newton's method) we need an equal number of equations and unknowns. *One way* of achieving this is to impose formally boundary conditions on the pressure, but to compute these conditions from eqs. (2.18)–(2.21). This is the way we chose, after having considered several other possibilities. In all other cases it was much more difficult to get the correct number of 'conditions' without relating the number of modes N to the number of grid points M.

Thus we compute 'boundary conditions' for the pressure. This again could be done in several different ways, using (2.21) with Dirichlet conditions (prescription of boundary values) or (2.18) with Neumann conditions (prescription of normal derivatives). Both versions were implemented into the program, and their numerical accuracy was compared for Couette flow. We found that (2.18) gave a better approximation of the pressure. Also, the resulting expressions were simpler. Thus only this version is discussed here.

Derivation of (2.19) w.r.t. r gives

$$-u''_\nu = \frac{1}{r}u'_\nu - \frac{1}{r^2}u_\nu + \nu k w'_\nu \qquad \nu = 1, \ldots, N \tag{2.26}$$

Substituting this into the second equation of (2.18) gives

$$p'_\nu + (\nu k)^2 u_\nu + \nu k w'_\nu = \text{Re } f_1(\mathbf{u}, \mathbf{u}', \mathbf{v}, \mathbf{w}; k; \nu, r). \tag{2.27}$$

Not to destroy the $\mathcal{O}(h^2)$-accuracy of our difference scheme, we evaluate (2.27) in $r_1 + h/2$ and in $r_2 - h/2$ using (2.23) with $h/2$ for first derivatives and expressions like

$$u(r_1 + \frac{h}{2}) = \frac{1}{2}(u(r_1) + u(r_1 + h)). \tag{2.28}$$

Concerning the evaluation of f_1, the numerical error is about the same when using (2.28) for the arguments of f_1 (implicit midpoint rule) or the trapezoidal rule

$$f_1(\ldots, r_1 + \frac{h}{2}) = \frac{1}{2}\left(f_1(\ldots, r_1) + f_1(\ldots, r_1 + h)\right). \tag{2.29}$$

The trapezoidal rule, however, gives a wider band of the Jacobian because it does not allow to use (2.23) with $h/2$ for the first derivatives in f_1. We thus used the implicit midpoint rule. In our computations for wide ranges of parameters we never had any problems with numerical stability.

2.2.4 Final version of the equations

The system of equations reads now, in detail:

$$
\begin{aligned}
p_o(r_1+\frac{h}{2}) &= 0,\\
\left(\nu k w'_\nu + (\nu k)^2 u_\nu + p'_\nu\right)[r_1+\frac{h}{2}] &= \mathrm{Re}\ f_1(\mathbf{u},\mathbf{u}',\mathbf{v},\mathbf{w};k;\nu,r_1+\frac{h}{2}),\\
p'_o(r) &= \mathrm{Re}\ f_1(\mathbf{u},\mathbf{u}',\mathbf{v},\mathbf{w};k;0,r),\\
\left(-u''_\nu - \frac{1}{r}u'_\nu + ((\nu k)^2+\frac{1}{r^2})u_\nu + p'_\nu\right)[r] &= \mathrm{Re}\ f_1(\mathbf{u},\mathbf{u}',\mathbf{v},\mathbf{w};k;\nu,r),\\
\left(u'_\nu + \frac{1}{r}u_\nu + \nu k w_\nu\right)[r] &= 0, \qquad\qquad (2.30)\\
\left(-v''_o - \frac{1}{r}v'_o + \frac{1}{r^2}v_o\right)[r] &= \mathrm{Re}\ f_2(\mathbf{u},\mathbf{v},\mathbf{v}',\mathbf{w};k;0,r),\\
\left(-v''_\nu - \frac{1}{r}v'_\nu + ((\nu k)^2+\frac{1}{r^2})v_\nu\right)[r] &= \mathrm{Re}\ f_2(\mathbf{u},\mathbf{v},\mathbf{v}',\mathbf{w};k;\nu,r),\\
\left(-w''_\nu - \frac{1}{r}w'_\nu + (\nu k)^2 w_\nu - \nu k p_\nu\right)[r] &= \mathrm{Re}\ f_3(\mathbf{u},\mathbf{v},\mathbf{w},\mathbf{w}';k;\nu,r),\\
r &= 1+\mu h,\ \mu = 1,\dots,M;\\
p'_o(r_2-\frac{h}{2}) &= \mathrm{Re}\ f_1(\mathbf{u},\mathbf{u}',\mathbf{v},\mathbf{w};k;0,r_2-\frac{h}{2}),\\
\left(\nu k w'_\nu + (\nu k)^2 u_\nu + p'_\nu\right)[r_2-\frac{h}{2}] &= \mathrm{Re}\ f_1(\mathbf{u},\mathbf{u}',\mathbf{v},\mathbf{w};k;\nu,r_2-\frac{h}{2}),\\
\nu &= 1,\dots N;
\end{aligned}
$$

with f_i as given in (2.22). We repeat them here for the convenience of the reader:

$$
f_1(\mathbf{u},\mathbf{u}',\mathbf{v},\mathbf{w};k;0,r) = \frac{1}{r}v_o^2 - \frac{1}{2}\sum_{j=1}^{N}(u_j u'_j - \frac{1}{r}v_j^2 - jku_jw_j), \qquad (2.31)
$$

$$
\begin{aligned}
&f_1(\mathbf{u},\mathbf{u}',\mathbf{v},\mathbf{w};k;\nu,r) =\\
&\quad \frac{2}{r}v_o v_\nu - \frac{1}{2}\left(\sum_{j=1}^{\nu-1}(u_j u'_{\nu-j} - \frac{1}{r}v_j v_{\nu-j} + (\nu-j)ku_{\nu-j}w_j)\right.\\
&\quad \left. + \sum_{j=\nu+1}^{N}(u_j u'_{j-\nu} + u'_j u_{j-\nu} - \frac{2}{r}v_j v_{j-\nu} - (j-\nu)ku_{j-\nu}w_j - jku_jw_{j-\nu})\right),
\end{aligned}
$$

$$f_2(\mathbf{u},\mathbf{v},\mathbf{v}',\mathbf{w};k;0,r) = -\frac{1}{2}\sum_{j=1}^{N}(u_j(v_j' + \frac{1}{r}v_j) - jkv_jw_j),$$

$$f_2(\mathbf{u},\mathbf{v},\mathbf{v}',\mathbf{w};k;\nu,r) =$$

$$- u_\nu v_o' - \frac{1}{r}u_\nu v_o - \frac{1}{2}\left(\sum_{j=1}^{\nu-1}(u_jv_{\nu-j}' + \frac{1}{r}u_jv_{\nu-j} + (\nu-j)kv_{\nu-j}w_j)\right.$$

$$+ \sum_{j=\nu+1}^{N}(u_jv_{j-\nu}' + v_j'u_{j-\nu} + \frac{1}{r}(v_ju_{j-\nu} + u_jv_{j-\nu})$$

$$\left. - (j-\nu)kv_{j-\nu}w_j - jkv_jw_{j-\nu})\right),$$

$$f_3(\mathbf{u},\mathbf{w},\mathbf{w}';k;\nu,r) =$$

$$-\frac{1}{2}\left(\sum_{j=1}^{\nu-1}(u_jw_{\nu-j}' + jkw_{\nu-j}w_j)\right.$$

$$\left. + \sum_{j=\nu+1}^{N}(w_j'u_{j-\nu} - u_jw_{j-\nu}' - \nu kw_{j-\nu}w_j)\right).$$

If r coincides with a grid point, all functions u_j, u_j', u_j'' are evaluated at this point using centered differences according to formulas (2.23) and (2.24). If r is not a grid point (first and last eqs. of (2.30)), we use the proper value of $1/r$ with (2.28) and the implicit midpoint rule, as explained in the previous section. An example is given in (2.52).

We can write (2.30) as

$$\mathbf{F}(\mathbf{x}, \mathrm{Re}, k, \eta, \mu) = A(k,\eta)\mathbf{x} - \mathrm{Re}\ \mathbf{f}(\mathbf{x};k,\eta) - \mathbf{b}(\eta,\mu) = 0, \tag{2.32}$$

which was already mentioned in (2.6). The quantities A, $\mathbf{x}$ and $\mathbf{f}$ are constructed according to the sequence of equations (2.30). The first elements of $\mathbf{x}$ are the p_ν at $r = r_1 + h/2$, followed by $p_\nu, u_\nu, v_\nu, w_\nu$ at $r = r_1 + h, \ldots$. The last elements are the p_ν at $r = r_2 - h/2$. Matrix A is block tridiagonal and of order $K := (2N+1)(2M+1)$. Vector $\mathbf{b}$ is generated by the inhomogeneous boundary conditions (2.3c):

$$\mathbf{b} = \mathbf{0} + b_{3N+2}(\eta) + b_{K-3N-1}(\eta,\mu). \tag{2.33}$$

The dependence of $\mathbf{b}$ on μ is explicit; on η it is both explicit and implicit. The dependence of A and $\mathbf{f}$ on k is explicit. Their dependence on η is implicit, through their dependence on r.

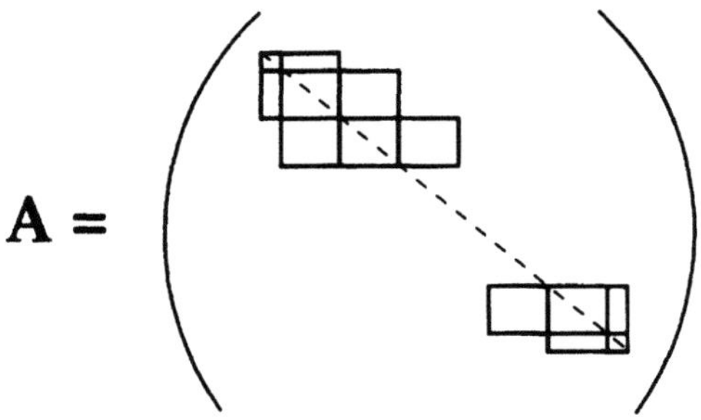

Figure 2.2: Sketch of matrix A.

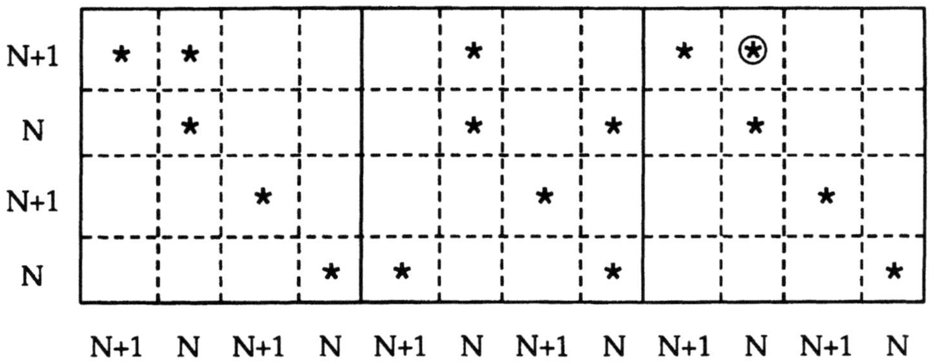

Figure 2.3: Sketch of A's block structure.

Fourier decomposition generates full matrices, finite differences generate block tridiagonal matrices. Thus we get a block tridiagonal matrix, whose band width strongly depends on the ordering of the equations and unknowns. With the ordering sketched in (2.30), the blocks of A have the structure shown in Fig. 2.3. The stars mark sub-matrices with non-zero elements. Most of them are sparse again. Considered as a band matrix, A has bandwidth $9N + 5$. But the band is unsymmetric w.r.t. the main diagonal.

Computation of the Jacobian causes fill-in only inside the band and in the sub-matrix $\circ$, which becomes full. The Jacobian of system (2.32) has band width $11N + 5$. Since our Gauss elimination (with column pivoting) required a band symmetric to the main diagonal, we worked with band widths $10N + 5$ and $12N + 5$. Altogether there is so much fill-in that the band of the Jacobian contains about 1/3 non-zero elements, 2/3rds zeroes. Efficient usage of software for sparse matrices requires $1/10^{th}$ or less non-zero elements in the band. In this sense, the Jacobian is not sparse. With a different ordering of the equations or unknowns, the band width of A could be considerably larger. We minimized it by methods given in [Tew].

2.3 Computation of solutions

Now we discuss the computation of solutions of the discretized system (2.6) for many different values of the parameters, and the computational side of investigating the bifurcation structure of the solution set.

One of the main problems in applying Newton's method is finding good initial values. When we vary one of the natural parameters in the problem monotonically, say β, the solution $\mathbf{x} = \mathbf{x}(\beta)$ varies smoothly, and we always have 'good' initial estimates for the Newton-Gauss iterations. This is accomplished by the method of continuation. Additionally, this method allows us to locate folds and bifurcation points and to switch branches at simple bifurcation points.

In this section we first describe the method of pseudo-arclength continuation for a general equation

$$G(u, \beta) = A(\beta)u - g(u, \beta) = 0 \tag{2.34}$$

containing a parameter β to be varied. Then we discuss special aspects of applying the method, and how to switch branches at simple bifurcation points. For a theoretical explanation and justification of the method see [Ke77, Ke78, Ke87].

2.3.1 Pseudo-arclength continuation and Newton iterations

In pseudo-arclength continuation with respect to the parameter β, we parameterize a family of solutions (u, β) of (2.34) by an additional parameter s. Introducing this new parameter allows us to introduce an additional scalar constraint. We do this in a manner such that s is an approximation to 'arclength'. Thus we now seek to solve

$$\begin{aligned} G(u(s), \beta(s)) &= 0, \qquad\qquad (2.35)\\ N(u, \beta, s) &= \theta < \dot{u}(s_o), [u(s) - u(s_o)] > \\ &\quad +(1-\theta)\dot{\beta}(s_o)[\beta(s) - \beta(s_o)] - (s - s_o) = 0. \end{aligned}$$

Here $(u(s_o), \beta(s_o))$ is a previously computed solution and $\dot{u} \equiv \frac{du}{ds}$, $\dot{\beta} = \frac{d\beta}{ds}$. Usually $\theta = 1/2$. $< u_1, u_2 >$ is a scalar product.

The Jacobian of the nonlinear system (2.35) is

$$C(s) = \begin{pmatrix} G_u(s) & G_\beta(s) \\ N_u(s) & N_\beta(s) \end{pmatrix}. \tag{2.36}$$

Computation of $\dot{u}(s_o)$, $\dot{\beta}(s_o)$: If $u(s_o), \beta(s_o)$ are known and the sub-Jacobian $G_u(s_o)$ is nonsingular, then we can compute $\dot{u}(s_o)$ and $\dot{\beta}(s_o)$ by solving

$$\begin{aligned} G_u(s_o)u_\beta &= -G_\beta(s_o), \\ \dot{\beta}(s_o) &= \pm[1+ < u_\beta, u_\beta >]^{-\frac{1}{2}}, \\ \dot{u}(s_o) &= \dot{\beta}(s_o)u_\beta. \end{aligned} \tag{2.37}$$

These relations come from the exact arclength condition

$$< \dot{u}, \dot{u} > + |\dot{\beta}|^2 = 1 \tag{2.38}$$

and the chain rule $G_u\dot{u} + G_\beta\dot{\beta} = 0$ for (2.34). The sign in (2.37b) is chosen such that continuation with a uniform orientation is insured (not to retrace previously computed paths). Solving (2.37) is theoretically equivalent to solving

$$C(s_o) \begin{pmatrix} \dot{u}(s_o) \\ \dot{\beta}(s_o) \end{pmatrix} = \begin{pmatrix} 0 \\ -N_s(s_o) \end{pmatrix} \tag{2.39}$$

by the *bordering algorithm.* If we solve version (2.39) by a Gauss elimination subroutine, we have to treat $C(s_o)$ as a full matrix. That is undesirable since the sub-matrix $G_u(s_o)$ is a band matrix, see the text near Figure 2.2. Version (2.37), however, allows us to take full advantage of the band structure of $G_u(s_o)$.

Bordering Algorithm: Any non-singular $n \times n$ linear system

$$\begin{pmatrix} B & b \\ c^* & d \end{pmatrix} \begin{pmatrix} p \\ q \end{pmatrix} = \begin{pmatrix} r \\ t \end{pmatrix}, \tag{2.40}$$

where B is a nonsingular $(n-1) \times (n-1)$ matrix and d is a scalar, can be solved by the following algorithm [Ke77]:

$$\begin{array}{llll} a) & By = b, & c) & q = (t - c^*z)/(d - c^*y), \\ b) & Bz = r, & d) & p = z - qy. \end{array} \tag{2.41}$$

The non-singularity of $C(s)$ ensures that $d - c^*y \neq 0$.

Newton-Gauss method: Now we take

$$\begin{aligned} u^o(s) &= u(s_o) + (s - s_o)\dot{u}(s_o) \\ \beta^o(s) &= \beta(s_o) + (s - s_o)\dot{\beta}(s_o) \end{aligned} \tag{2.42}$$

as initial guess for Newton's method

$$C(\nu, s)\begin{pmatrix} u^{\nu+1} - u^\nu \\ \beta^{\nu+1} - \beta^\nu \end{pmatrix} = -\begin{pmatrix} G(u^\nu, \beta^\nu) \\ N(u^\nu, \beta^\nu, s) \end{pmatrix}. \tag{2.43}$$

This system is solved by the bordering algorithm, and (2.41a) and (2.41b) are solved simultaneously by Gauss elimination with partial pivoting. As long as we had not implemented a step size control, we used the following strategy: After

$$\max |(u^\nu, \beta^\nu) - (u^{\nu-1}, \beta^{\nu-1})| \leq 10^{-3} \tag{2.44}$$

was satisfied, we switched to the simplified Newton method. That saves the time both for computing new Jacobians G_u and for doing more *LU*-decompositions. The two back substitutions for solving (2.41a) and (2.41b) take very little time. In most cases the convergence continued to be faster than linear, until we got the 'solution', i.e. until the following two conditions were satisfied:

$$\max |(u^\nu, \beta^\nu) - (u^{\nu-1}, \beta^{\nu-1})| \leq 10^{-7}, \tag{2.45}$$

$$\max |G(u^\nu, \beta^\nu)| = \max |A(\beta^\nu)u^\nu - g(u^\nu, \beta^\nu)| \leq 10^{-7}. \tag{2.46}$$

A check of the necessary condition (2.45) is standard, but a check of the sufficient condition (2.46) is highly recommended. Checking (2.46) is not much extra work since $G(u^\nu, \beta^\nu)$ is part of the r.h.s. of (2.43) and is thus available at each step. In the beginning, only (2.45) was used to terminate the iterations, but $\|G(u^\nu, \beta^\nu)\|$ was watched. (2.46) used to be satisfied whenever (2.45) was. After some substantial changes in the program, however, a discrepancy could be seen: (2.45) was satisfied, but in (2.46) a factor of 10^{-4} was still missing. We thus implemented (2.46) as an additional condition for the termination of the iterations. This turned out to be very valuable when we discovered that the increase of the necessary number of iterations was caused by a programming error. This error affected the computation of the Jacobian with respect to the wave number, but it did not affect the computation of the iterated matrix $A(\beta^\nu)$ of (2.34). Thus we did not perform proper Newton iterations and this slowed down the convergence, but the residuals computed were correct. Satisfying (2.46) thus saved us the work of redoing all the computations that had been performed with the wrong Jacobian.

Step Size Control: The size of the step $s - s_o$ in (2.42) has to be chosen with care: If it is taken too large, there might be convergence problems of the Newton iterations in a good case, uncontrolled branch switching in a bad case. If it is taken too small, the computations become inefficient. In the appropriate range, Newton's method converges monotonically and quadratically. Within this appropriate range, there is a trade-off between the number of steps in the parameter and the number of Newton iterations.

Quite sophisticated algorithms for step size control can be found in the literature. A simple to use, robust step size control is the following [WS87]: Chose the desired number I of (proper) Newton iterations. Let I_1 be the number of iterations actually needed for the most recent solution attained with step-size $t_1 := s - s_o$. Replace $s - s_o$ in (2.42) by $t_1 I/I_1$ to find the new initial values for the next Newton iterations.

This worked very well. In all cases the number of iterations needed to satisfy the convergence criteria were found to be a good measure of how much the solutions change with the parameter. In most cases we easily found values for I that ensured quadratic convergence of all Newton iterations in large parameter ranges. In these cases the switching between proper Newton and simplified Newton seemed not so important.

2.3.2 Continuation in the Reynolds number Re

In this section we discuss how to use the method of pseudo-arclength continuation with respect to $\beta = \text{Re}$. In this case the parameter dependent nonlinear system reads

$$G(u, \beta) = Au - \beta f(u) + b. \tag{2.47}$$

As scalar product we chose

$$< u_1, u_2 >= c\, h^2 u_1^* \cdot u_2, \tag{2.48}$$

where h is the mesh spacing and c is a constant that had been chosen experimentally. The change of the scalar product from $< u_1, u_2 >= h^2 u_1^* \cdot u_2$ to (2.48) essentially made h independent of the gap width between the cylinders. After that change (that can be interpreted as a change in θ of (2.35)), start function and converged solution were much closer together.

That was especially important in the branch-switching process. In later work [BoKe] an approximation of

$$< u_1, u_2 >= \frac{\int u_1(r,z)u_2(r,z)rdrdz}{\int rdrdz} \tag{2.49}$$

was used, with analytically integrated denominator and numerator by the trapezoid rule. This has the property that the norm of the solution bifurcating from Couette flow is independent of the number of cells.

The main work in computing the Jacobian $C(s)$ of (2.36) is to compute

$$G_u(u,\beta;s) = A - \beta f_u(u), \tag{2.50}$$

and hence to compute f_u. And this has to be done quite often for (2.43), and thus in an efficient way. So we have to look again at eqs. (2.31). Take the first one,

$$f_1(\mathbf{u},\mathbf{u}',\mathbf{v},\mathbf{w};k;0,r) = \frac{1}{r}{v_o}^2 - \frac{1}{2}\sum_{j=1}^{N}(u_j u_j' - \frac{1}{r}{v_j}^2 - jku_j w_j). \tag{2.51}$$

After discretization, it reads

$$\begin{aligned} f_1(\mathbf{u},\mathbf{u}',\mathbf{v},\mathbf{w};k;0,\mu) &= \frac{1}{r_1+\mu h}{v_o}^2(r_1+\mu h) + \\ -\frac{1}{2}\sum_{j=1}^{N}[u_j(r_1+\mu h)&\frac{u_j(r_1+(\mu+1)h)-u_j(r_1+(\mu-1)h)}{2h} + \\ -\frac{1}{r_1+\mu h}{v_j}^2(r_1+\mu h) &- jku_j(r_1+\mu h)w_j(r_1+\mu h)]. \end{aligned} \tag{2.52}$$

Now we have to compute the derivatives of this function with respect to each

$$(p_j(\mu), u_j(\mu), v_j(\mu), w_j(\mu))_{j=0,1,\ldots,N;\ \mu=1,\ldots,M}\,. \tag{2.53}$$

Taking advantage of the very special structure of the function f, we put

$$\frac{\partial\, f^{\nu,\mu}}{\partial u_j(\mu)} = f^{\nu,\mu}|_{u_j(\mu)=1} - f^{\nu,\mu}|_{u_j(\mu)=0}. \tag{2.54}$$

That is exact for all variables that occur only linearly and was overwritten with the correct derivatives for all quadratically occurring variables, in the case of (2.52) for $v_j(\mu)$, $j=0,\ldots,N$, $\mu=1,\ldots,M$.

For testing and debugging we compared this Jacobian band matrix with the full discrete Jacobian containing

$$\frac{\partial f^{\nu,\mu}}{\partial u_j(\mu)} = \frac{f^{\nu,\mu}(u_j(\mu)+d) - f^{\nu,\mu}(u_j(\mu))}{d}. \tag{2.55}$$

Numerically, they were nearly equivalent for $d = 10^{-2}$, since the nonlinearities are all quadratic. Already for $N = 2,\ M = 3$ (35 unknowns), Approach I, using (2.54) and computing only non-zero entries, needed roughly half as much computing time as Approach II, using (2.55) and computing all elements of the matrix [MK78, p. 27f]:

1× Approach I, 1× Approach II: 2 sec;
200× Approach I, 1× Approach II: 32 sec;
1× Approach I, 200× Approach II: 64 sec on an IBM 370/158.

2.3.3 Continuation in the wave number k

In this section we discuss how to use the method of pseudo-arclength continuation with respect to $\beta = k = \frac{2\pi}{\lambda\delta}$. The parameter dependent nonlinear system to be solved now reads

$$G(u,\beta) = A(\beta)u - \text{Re}\ f(u,\beta) + b. \tag{2.56}$$

The main point is how to compute the Jacobian $C(\nu, s)$ of (2.36) in this case. We have to compute

$$G_u(u^\nu,\beta^\nu;s) = A(\beta^\nu) - \text{Re}\ f_u(u^\nu,\beta^\nu) \tag{2.57}$$

and

$$G_\beta(u^\nu,\beta^\nu;s) = A_\beta(\beta^\nu)u^\nu - \text{Re}\ f_\beta(u^\nu,\beta^\nu). \tag{2.58}$$

The computation of (2.57) for given $\beta^\nu = k$ and Re was already needed for the arclength continuation in Re and is thus available (see the previous section for details). In addition the previously constant A has now to be recomputed in each iteration step. From eqs. (2.30) and (2.31) we see that the entries of A depend at most quadratically on $k = \beta^\nu$, while the dependence of the entries of the vector f is linear. Thus

$$f_\beta(u,\beta) = \frac{f(u,\beta+\varepsilon) - f(u,\beta)}{\varepsilon} \tag{2.59}$$

and

$$A_\beta(\beta) \approx \frac{A(\beta+\varepsilon) - A(\beta-\varepsilon)}{2\varepsilon} \tag{2.60}$$

for any reasonably sized $\varepsilon > 0$. We compute f_β according to (2.59). Instead of (2.60), however, we use

$$A_\beta(\beta)u \approx \frac{1}{\varepsilon}\left(A(\beta+\varepsilon)u - A(\beta)u\right). \tag{2.61}$$

The use of the one-sided formula saves the computer time needed to replace the already computed $A(\beta)$ by $A(\beta-\varepsilon)$, and the early multiplication by u saves the storage space of an additional matrix. Numerical tests with various ε's showed that the error introduced by the unsymmetric formula is negligible. With $\varepsilon = 10^{-5}$ we usually observed quadratic convergence of the Newton iterates.

2.3.4 Simple continuation

The computation of the tangent $(\dot{u}, \dot{\beta})[s_o]$ in (2.37) is some extra work, worthwhile because it allows us to pass fold points and because it provides the information needed for switching branches at simple bifurcation points. If the circumstances are such that Newton's method easily converges, it is more efficient to compute the solutions by simple continuation. This is the case when the solutions on a large portion of a branch are well isolated (i.e. have a large domain of attraction around them), away from the singularities of a branch.

Suppose we have a solution (u_o, β_o) of (2.34) and we want to compute a solution (u_1, β_1), where $\beta_1 = \mathrm{Re}$ or $\beta_1 = k$ is given and all other parameters are assumed to be fixed. In simple continuation, we use (u_o, β_1) as initial guess for Newton iterations with the Jacobian $G_u(u^\nu, \beta_1)$. The choice of the step size $\beta_1 - \beta_o$, switching to simplified Newton iterations and termination of the iterations is done in a similar way as in the pseudo-arclength iterations.

Convergence? If (u_o, β_o) is an isolated solution of (2.34), i.e. if the Jacobian $G_u(u_o, \beta_o)$ is nonsingular, then for $|\beta_o - \beta_1|$ small enough, there is an isolated solution (u_1, β_1) belonging to the same branch as (u_o, β_o), and (u_o, β_1) belongs to its domain of attraction. This follows from the Implicit Function Theorem [Me75].

Moving along a branch of solutions is optimal for obtaining convergence. Even when the domain of attraction is shrinking because a singular point is approached, it is possible to 'reach' that point by simple continuation in the accuracy of the computer used. In that case, however, the successful step size gets unreasonably small: pseudo-arclength continuation allows for larger steps and is thus more efficient.

Examples:
a) In [Me75], the fold point of

$$\Delta u = -\lambda e^u \quad \text{in } D := [0,1] \times [0,1], \qquad u|_{\partial D} = 0 \tag{2.62}$$

was approached by simple continuation from both sides up to a difference less than 10^{-8} in the two solutions and in the parameter. Then the computations were stopped because it seemed unreasonable to continue (computer accuracy $\approx 10^{-15}$, discretization error larger than 10^{-8}).
b) Even when the structure of the solution set is quite complicated but known, and when it is easy to find a starting function on each of the branches, it is possible to compute the whole solution diagram by simple continuation. Riecke and Paap computed the branches displayed in Fig. 3.36 by simple continuation, knowing the structure of the solution set from Fig. 3.4. For obtaining Fig. 3.4 for the first time, however, it seemed necessary to implement pseudo-arclength continuation with respect to the wave number k ([FM] and section 'continuous change of period' in this text).$\diamond$

When using simple continuation for fixed Reynolds number and varying wave number k, we found the necessary size of the steps in k to depend strongly on the direction in which k was varied: approaching a fold point or a bifurcation point usually demands a much smaller step size than leaving it.

2.3.5 Switching branches

When continuing along solution branches, we encountered many bifurcation points; simple ones and multiple ones, pitchfork, transcritical, subcritical, ... Here we consider only the simplest, generic case: two smooth branches have a non-tangential intersection. And we deal with the following question: If we know a solution and the tangent on one branch, how do we compute tangent and starting values for the other branch?

Though there are so many multiple bifurcations in the problem, and though they occur with continuation in any of the parameters, we found it sufficient to implement the branch switching algorithm at a simple bifurcation point with respect to one of the parameters, i.e. the Reynolds number.

The combination of this branch switching algorithm with pseudo-arclength continuation in the other parameters allowed us the investigation of all solution branches we were interested in, even in the involved cases reported near Fig. 3.11 and Fig. 3.26. Maybe this is so because the Taylor problem has an inherent asymmetry and thus so many bifurcations are perturbed.

Several algorithms are known to switch from one solution branch of

$$G(u, \beta) = 0 \tag{2.63}$$

to the other. We chose method I of [Ke77, sections 5, 6] and changed it slightly. It seemed to be the simplest one for the problem under consideration. If we already know that we are close to a bifurcation point, it consists of the following steps:

1) Locate the bifurcation point (u_o, β_o) by finding β_o such that $G_u^o := G_u(u_o, \beta_o)$ is singular.
   Note: when we execute the following formulas, we use a numerical approximation to the singular matrix. Hence G_u^o is typically nonsingular, with very large condition number.
2) Find ϕ_1, ψ_1 with

$$a)\ G_u^o \phi_1 \stackrel{!}{=} 0, \qquad b)\ G_u^{o*} \psi_1 \stackrel{!}{=} 0, \qquad c)\ < \psi_1, \phi_1 >= 1 \tag{2.64}$$

   by Inverse Iteration (which converges very fast, because of the large condition number).
3) Solve (2.37a) for $\phi = u_\beta$.
4) Compute

$$a)\ \alpha =< \psi_1, \phi > \qquad \text{and } b)\ \phi_o = \phi - \alpha \phi_1. \tag{2.65}$$

5) Compute

$$\begin{aligned} a_{111}(\varepsilon) &= < \psi_1, \frac{1}{\varepsilon}\{G_u(u_o + \varepsilon\phi_1, \beta_o) - G_u^o\}\phi_1 > \qquad (2.66) \\ b_{11}(\varepsilon) &= < \psi_1, \frac{1}{\varepsilon}\{G_u(u_o + \varepsilon\phi_1, \beta_o) - G_u^o\}\phi_o + \\ &\quad + \frac{1}{\varepsilon}\{G_\beta(u_o + \varepsilon\phi_1, \beta_o) - G_\beta^o\} > \end{aligned}$$

6) Compute

$$\begin{aligned} &a)\quad \tilde{\alpha} = -(\alpha + \frac{2b_{11}(\varepsilon)}{a_{111}(\varepsilon)}) \\ &b)\quad \bar{\alpha_o} = \pm 1/(1+\tilde{\alpha}^2)^{\frac{1}{2}} \qquad (2.67) \\ &c)\quad \bar{\alpha_1} = \tilde{\alpha} \cdot \bar{\alpha_o} \end{aligned}$$

7) Then $(\dot{u}_2, \dot{\beta}_2)$, computed by

$$a)\ \dot{u}_2 = \bar{\alpha}_1 \phi_1 + \bar{\alpha}_o \phi_o \qquad b)\ \dot{\beta}_2 = \bar{\alpha}_o \tag{2.68}$$

is the tangent of the bifurcated branch at (u_o, β_o).

8) Take

$$u^o = u_o + (s - s_o)\dot{u}_2, \qquad \beta^o := \beta_o + (s - s_o)\dot{\beta}_2 \tag{2.69}$$

as initial guess to compute a solution on the bifurcated branch by Newton's method.

We comment on some of the steps in this algorithm.

Locating the bifurcation point: We assumed the bifurcation point to be a non-tangential intersection of two smooth branches. Thus we can locate it by searching for a sign change of the Jacobian on the current branch. If we search for a bifurcation point on Couette flow, we increase Re in fixed step-length until the first sign change occurs. If we compute along a bifurcated branch, we monitor the sign of the product of the pivots in the LU-decomposition during the Newton iterations and use the Re-values of adjacent converged solutions.

After we found Re_1 and Re_2 with

$$\det G_u(\mathbf{x}(\mathrm{Re}_1), \mathrm{Re}_1) \cdot \det G_u(\mathbf{x}(\mathrm{Re}_2), \mathrm{Re}_2) < 0, \tag{2.70}$$

we iterate for the zero until either

$$a)\ \frac{|\mathrm{Re}_2 - \mathrm{Re}_1|}{\mathrm{Re}_1} \leq 10^{-8} \quad \text{or} \quad b)\ \text{pivmin} \leq 10^{-8}, \tag{2.71}$$

pivmin being the minimum absolute value of a pivot in the LU-decomposition of the Jacobian. This iteration can be done by bisection or false position on $\det G_u(\mathrm{Re})$. We preferred to work with pivmin $\mathrm{sgn}(\det G_u(\mathrm{Re}))$ instead: already for $N = 2$, $M = 3$ and $|\mathrm{Re}_2 - \mathrm{Re}_1|/\mathrm{Re}_i \approx 4 \cdot 10^{-5}$, $i = 1, 2$ we found a huge determinant: $\log_{10} |\det G_u(\mathrm{Re}_i)| > 95$ for the 2-vortex bifurcation from Couette flow at moderate wavelength λ. Though the determinant depends smoothly on the parameters, we found that, near its sign change, it could be well approximated by a jump of a Heaviside function. The minimum pivot changes its absolute value pivmin less abruptly, and it carries the relevant information at single zeroes.

Formula (2.65): Since the solutions on the first branch are known, we eliminated α_o and α_1 from the formulas given in [Ke77, section 6], saving thus some divisions and multiplications.

Formula (2.66): Since the Jacobian G_u depends linearly on all its variables, we choose to approximate G_{uu} and $G_{\beta\beta}$ by their discrete analogs. A test run showed that $a_{111}(\varepsilon)$ and $b_{11}(\varepsilon)$ are indeed completely independent of the choice of ε.

Formula (2.67): This is an algorithm to solve

$$a)\ \bar{\alpha_1}/\bar{\alpha_o} = -(\alpha + \frac{2b_{11}}{a_{111}}), \qquad b)\ \bar{\alpha_o}^2 + \bar{\alpha_1}^2 = 1. \tag{2.72}$$

The sign in (2.67b) is arbitrary and determines on which half of the bifurcated branch the computations proceed. To insure that we continue to proceed in the same direction, away from the bifurcation point, the sign should be chosen such that two neighboring tangent vectors make an acute angle with each other. This can be assured by requiring positivity of the inner product of the two tangents [Ke78, (4.8)]. If this condition is not imposed the computations may be very sensitive to small changes in the numerical approximation of the null vectors ϕ_1, ψ_1 and of Re_{bif}.

Example: For the values $R_1 = 1.9$, $R_2 = 2$, $\eta = 0.95$, $\lambda = 2.007$, $N = 2$, $M = 31$ we obtained in early computations the following results:

Re_{bif}	pivmin	$< \psi_1, G^o_\lambda >$	$\bar{\alpha_o}$	$\bar{\alpha_1}$	
3499.9113	$3 \cdot 10^{-5}$	$5 \cdot 10^{-9}$	$1.6 \cdot 10^{-9}$	1.	(2.73)
3499.9109	$2 \cdot 10^{-9}$	$3 \cdot 10^{-12}$	$2.5 \cdot 10^{-13}$	−1.	

Here Re_{bif} is the numerical approximation to the critical Reynolds number, $< \psi_1, G^o_\lambda > \approx 0$ indicates that we may be at a bifurcation point [Ke77, p. 378]. So if we do the branch switching process with both values for Re_{bif}, we get convergence to the two different parts of the bifurcating branch. The solutions on the two parts of the branch are called '$\lambda/2$ shifted twins' later on. One can be obtained from the other one by shift of half their wavelength λ. This is explained in detail in the section named 'Periodicity'. ◇

2.4 Computation of flow parameters

Having computed solutions, we want to use them for investigation and better understanding of flow phenomena to be discussed in the next two chapters. Also, we want to compare computed values with measurements from experiments. Thus various physical quantities have to be computed: torque, kinetic energies, stream function, radial velocity at the midpoint of the computational domain. Some of these are dimensional, some dimensionless. The Reynolds number Re relates the dimensional and dimensionless quantities. We recall definitions given near (2.2) and add others:

$$\begin{aligned} \mathrm{Re} &= \Omega_1 R_1^2/\nu, \qquad \mu = \Omega_2/\Omega_1, \qquad \delta = (R_2 - R_1)/R_1, \\ \eta &= R_1/R_2, \qquad k = 2\pi R_1/L, \qquad \ell = L/R_1, \end{aligned} \tag{2.74}$$

where R_1 (R_2) is the dimensional inner (outer) radius, $\Omega_i R_i$ are the velocities of the two cylinders, L is the dimensional period of the computational domain, and ℓ its dimensionless counterpart. Note that

$$\lambda = \ell^\delta = \frac{L}{R_2 - R_1} \tag{2.75}$$

is the dimensionless period corresponding to Re^δ defined in (1.10). The dimensional velocity components and radius are given by

$$U = \Omega_1 R_1 u, \quad V = \Omega_1 R_1 v, \quad W = \Omega_1 R_1 w; \qquad R = R_1 r. \tag{2.76}$$

Because of our normalizations, the dimensionless Couette flow does not depend on Re. Its stability, however, does.

As we have seen near Fig. 1.9, the value of the critical Reynolds number depends on the parameters in the problem, on η, μ and on λ. To make it easier to compare the various results, we use mostly the 'normalized Reynolds number' $\mathrm{Re}/\mathrm{Re}_{cr}$, where Re_{cr} is the minimum value in the marginal curve for given η and μ, as defined in (1.37). Note that

$$T/T_{cr} = (\mathrm{Re}/\mathrm{Re}_{cr})^2, \tag{2.77}$$

where T is any of the Taylor numbers preferred in perturbation approaches and T_{cr} is defined analogously to Re_{cr}; see (1.36) for how T arises.

2.4.1 Periodicity

The **representation** of solutions of (2.30) by expressions of the form (2.11) is **not unique**:
Let **a** be a solution of (2.30) with radial velocity

$$u_a(r, z) = \sum_{\nu=1}^{N} a_\nu(r) \cos \nu k_a z \tag{2.78}$$

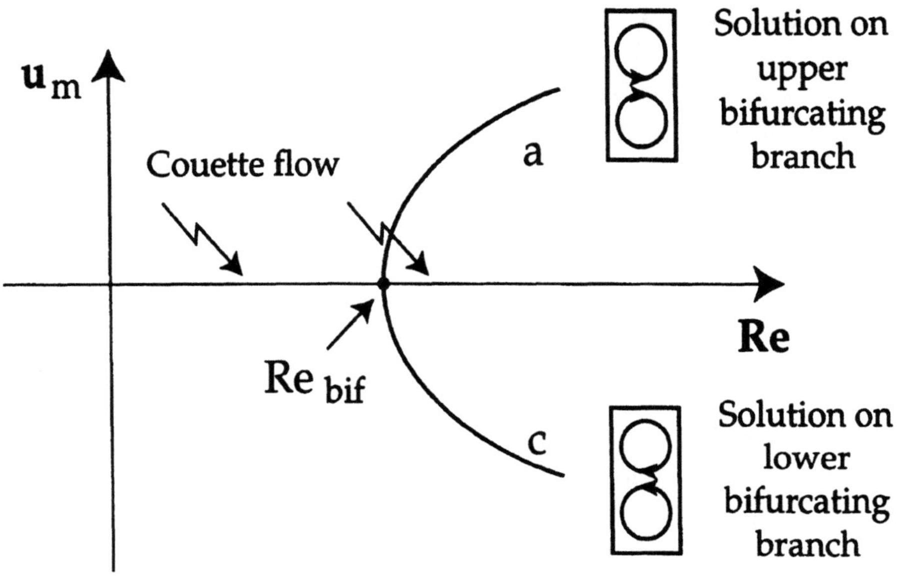

Figure 2.4: Bifurcation from Couette flow

and period $\lambda_a = 2\pi(k_a\delta)^{-1} = 2\pi(k_a D/R_1)^{-1}$. We define **b** by

$$u_b(r,z) := \sum_{\nu=1}^{N} a_\nu(r)\cos 2\nu\frac{k_a}{2}z = \sum_{\nu=1}^{2N} b_\nu(r)\cos\nu k_b z, \tag{2.79}$$

with $k_b := k_a/2,\ \lambda_b := 2\lambda_a$, and

$$b_\nu(r) := \begin{cases} 0 & \text{for } \nu \text{ odd} \\ a_{\nu/2}(r) & \text{for } \nu \text{ even} \end{cases}. \tag{2.80}$$

Then **a** *and* **b** *represent the same solution in infinitely long cylinders, considered 'through windows of different lengths'.* We call **b** the **periodic repetition** of **a**, or the $\times 2$-branch.

In the same way we can generate solutions with period $m\lambda_a$ with arbitrary $m \in \mathbb{N}$. The branch of such solutions is called the $\times$**m-branch**. Note that the computation of solutions on the $\times m$-branch requires *m-times as many* Fourier modes for achieving *the same* numerical accuracy as the computations on the $\times 1$-branch.

We got **b** from **a** by reinterpretation of the arguments of sin and cos. Now we consider **a shift of the solution** a **by half a wavelength**: $kz \mapsto kz+\pi =: k\tilde{z}$ and get a quantity **c**. This again is a solution of (2.30) because it also satisfies all boundary conditions. It also has the same symmetries with respect to $\tilde{z} = 0$ as **a** has w.r.t. $z = 0$. Its representation of the form (2.11) follows from

$$\begin{aligned}
\cos\nu(kz+\pi) &= \cos\nu\pi\cos\nu kz - \sin\nu\pi\sin\nu kz \\
&= (-1)^\nu \cos\nu kz; \qquad (2.81) \\
\sin\nu(kz+\pi) &= \sin\nu\pi\cos\nu kz + \cos\nu\pi\sin\nu kz \\
&= (-1)^\nu \sin\nu kz. \qquad (2.82)
\end{aligned}$$

a and **c** thus differ only in the signs of the coefficients of the odd modes. If we consider infinitely long cylinders, the two solutions are identical *in an abstract mathematical sense.* But *in every given coordinate system* these two solutions have different representations. Also, we wish to model long, but finite cylinders, with a real liquid between them. The boundary conditions (2.3f) don't allow the liquid to leave the computational array, and some forcing would be required to make the liquid move by half a wavelength. Actually, we need **a** and **c** together to get the familiar pitchfork bifurcation picture. In the following we call the solutions **a** and **c the** $\lambda/2$ **shifted twins.** Because of the forced broken symmetry of the Taylor problem discussed near Fig. 3.17, a big part of the bifurcation results to be discussed can only be understood if we consider both twin solutions simultaneously. (Only one of them was computed, the other one was generated according to formulas (2.81), (2.82)).

2.4.2 Computation of $u_m := u(r_m, z_m; \lambda)$ at $r_m := 1 + \delta/2,\ z_m = 0$

Let a solution **a** be computed, with radial velocity

$$u_a(r, z; \lambda_a) = \sum_{\nu=1}^{N} a_\nu(r) \cos\nu k_a z \qquad (2.83)$$

and wavelength $\lambda_a = (2\pi/k_a)(R_1/D)$. Then

$$u_a(r_m, 0; \lambda_a) = \sum_{\nu=1}^{N} a_\nu(r_m) \qquad (2.84)$$

$$\text{and} \quad u_c(r_m, 0; \lambda_a) = \sum_{\nu=1}^{N} (-1)^\nu a_\nu(r_m), \qquad (2.85)$$

where $\mathbf{c}(r,z) = \mathbf{a}(r, z + \pi/k_a)$ is the $\lambda/2$-shifted twin, see the previous subsection. For displaying the $\times m$-branches of **a** and **c** (called **b** in the previous subsection), no computations of velocity components are necessary: λ_a only needs to be replaced by $m\lambda_a$, $m = 2, 3, \dots$.

These same simplifications have been used in the figures showing other quantities (like toroidal velocity $v_m := v(r_m, z_m; \lambda)$ or kinetic energies) of periodic repetitions or shifted twins.

How this method of representation can be used for investigation of discretization errors is discussed in the section on numerical accuracy.

2.4.3 Computation of the torque

The torque $G(R)$ of a flow with azimuthal velocity $V(R, \Theta, Z)$ on a cylinder of length $L_o = L_2 - L_1$ is given by

$$G(R) = -\int_{L_1}^{L_2}\int_{0}^{2\pi} \nu\rho \left(\frac{\partial V}{\partial R} - \frac{V}{R}\right) R^2 \, dZ \, d\Theta \tag{2.86}$$

where

ν is the kinematic viscosity,
ρ is the density,
R is the dimensional radius as given in (2.76), $Z = R_1 z$, and
V is the dimensional velocity in the azimuthal direction (i.e. $V = v\,\Omega_1 R_1$).

For the Z and Θ independent Couette flow with azimuthal velocity $\mathcal{V}(R)$, (2.86) reduces to

$$G(R_2) = \frac{4\pi\nu\rho R_1 R_2^2 L_o \mathcal{V}(R_1)}{R_2^2 - R_1^2}, \tag{2.87}$$

as given in (1.14) and in [DS, p. 402].

For our computations, it is natural to compute the torque with the length being the period of the flow. In the experiments, however, the torque was measured in a certain fixed length L_0, irrespective of the period of the flow. We want to use the torque not only for comparing our computations with measurements, but also for comparing the physical properties of flows with different wavelengths. This is achieved by defining an **average torque**: we first compute the torque of the flow in one period, and then use that value to get the torque on a cylinder of given length ($L_0 = 5$ cm in the measurements of Donnelly/Simon [DS]). The average torque depends linearly on L_0.

Now we compute the torque of one period of a flow with dimensionless period ℓ and dimensional period L. Since

$$\int_{-\ell/2}^{\ell/2} v_\mu(r) \cos \mu k z \, dz = 0 \qquad \text{for } \mu \neq 0, \tag{2.88}$$

where $v_\mu(r)$ are the Fourier coefficients in (2.11), the torque is determined by the z-independent term in the Fourier series. Because of axisymmetry, (2.86) reduces further to

$$G(R) = -2\pi\nu\rho L R^2 \left(\frac{dV_0}{dR} - \frac{V_0}{R} \right), \tag{2.89}$$

with

$$V_0(R) = R_1 \Omega_1 \, v_o(r) = v_o(r) \text{ Re } \nu / R_1, \quad R = R_1 r. \tag{2.90}$$

We get the *average torque* of the flow on a cylinder of length L_0 by dividing the torque given by (2.89) by L and multiplying by L_0, where L_0 need not be an integer multiple of the dimensional wavelength.

We evaluate (2.90) at $r_1 + h/2$ and $r_2 - h/2$. As long as we are only considering steady states, we should get

$$G(R_1) = G(R_2). \tag{2.91}$$

Numerically, of course, the two differ. In all plots we give $G(R_1)$. We mostly found $G(R_2) > G(R_1)$, and $|G(R_2) - G(R_1)|$ always decreased when we refined the mesh [MK78, p. 35].

2.4.4 Computation of kinetic energies

The (dimensional) kinetic energy in an annulus of length $L_o = L_2 - L_1$ is given by

$$\begin{aligned} E_{Kin} &:= \int \frac{1}{2} \rho \mathbf{V}^2 R \, dR \, d\Theta \, dZ \qquad (2.92) \\ &= \pi\rho \int_{L_1}^{L_2} \int_{R_1}^{R_2} (U^2 + V^2 + W^2) R \, dR \, dZ \end{aligned}$$

Here
ρ is the density of the fluid,
Θ, R, Z are the dimensional cylindrical coordinates, and
$\mathbf{V} = (U, V, W)$ is the dimensional velocity.
Just as for the torque (previous subsection), we compute an **average** value of the **kinetic energy** in a volume with given length L_0. L_0 was chosen to be the same as before, $L_0 = 5$ cm. This allows us to compare the kinetic energies of solutions with different wavelengths.

There are no measurements of kinetic energies to compare with. It is thus sufficient and more convenient to consider relative kinetic energies

$$E_{Kin}/E_{Cr}, \tag{2.93}$$

where E_{Cr} was chosen to be the kinetic energy of Couette flow at $(\text{Re}_{cr}, \lambda_{cr})$. This way constant factors common to numerator and denominator need not be computed explicitly.

It is no extra effort to compute the kinetic energy of each velocity component separately and then add them. Thus we could study the parameter dependence of each velocity component u, v, and w separately. Other authors have done that [Mey67]. It does not make sense, however, to study the radial component u and the axial component w separately, because they are coupled by the equation of continuity

$$u_r + \frac{1}{r}u + w_z = 0. \tag{2.94}$$

This equation is satisfied identically by introduction of a **stream function** ψ satisfying eq. (2.8). To introduce ψ is possible because the $\mathbf{v}$-field is axisymmetric and divergence-free. A proof of existence for ψ uses the theorem giving the conditions for existence of a potential, Stokes' integral theorem and eq. (2.110). After introducing ψ we have

$$(u, v, w) = \left(\frac{1}{r}\frac{\partial \psi}{\partial z}, v, -\frac{1}{r}\frac{\partial \psi}{\partial r}\right). \tag{2.95}$$

That part of the flow which is given by ψ lies in the (r, z)-plane and is called the **poloidal flow**, while the φ-component v is called the **toroidal flow**. It does make sense to decompose the velocity field $\mathbf{v}$ into its toroidal and its poloidal component and to consider the toroidal component v and the poloidal (u, w)-field separately. This decomposition is standard for axisymmetric magnetic fields in magnetohydrodynamics.

The definitions are [Cha, Appendix III], [Mo78, p. 19]:

Let f and g be arbitrary smooth scalar functions of position $\mathbf{s} \in I\!R^3$. A vector field $\mathbf{t}$ is called **toroidal field** with **defining scalar** $f(\mathbf{s})$ if

$$\mathbf{t} = \text{curl}\left(\frac{f}{s}\mathbf{s}\right). \tag{2.96}$$

A vector field $\mathbf{q}$ is called **poloidal field** with **defining scalar** $g(\mathbf{s})$ if

$$\mathbf{q} = \text{curl}\left[\text{curl}\left(\frac{g}{s}\mathbf{s}\right)\right]. \tag{2.97}$$

It follows from basic rules of vector analysis that toroidal as well as poloidal fields are divergence-free, independent of the choice of f and g. Given a divergence-free vector field, the decomposition into toroidal and poloidal field is always possible [Mo78]. There are configurations (and boundary conditions), however, for which it is more practical to choose another $\mathbf{s}$ than the vector of position and to decompose the divergence-free vector field into (toroidal + poloidal + mean) flow [SW92].

We now assume that suitable functions f and g exist and evaluate formulas (2.96) and (2.97) in cylindrical coordinates (r, φ, z) with vector of position $\mathbf{s} := (r, 0, z)$ and distance $s := \sqrt{r^2 + z^2}$. We thus get

$$\begin{aligned} \mathbf{t} &= \text{curl}\left(\frac{f}{s}\mathbf{s}\right) = \text{curl}\left(\frac{fr}{s}, 0, \frac{fz}{s}\right), \\ &= \left(\frac{1}{r}\frac{\partial}{\partial\varphi}\frac{fz}{s} - 0, \frac{\partial}{\partial z}\frac{fr}{s} - \frac{\partial}{\partial r}\frac{fz}{s}, 0 - \frac{1}{r}\frac{\partial}{\partial\varphi}\frac{fr}{s}\right), \\ &= \left(\frac{1}{r}\frac{z}{s}\frac{\partial f}{\partial\varphi}, \frac{r}{s}\frac{\partial f}{\partial z} - \frac{z}{s}\frac{\partial f}{\partial r} - \frac{frz}{s^3} + \frac{frz}{s^3}, -\frac{1}{s}\frac{\partial f}{\partial\varphi}\right), \\ &= \left(0, \frac{r}{s}\frac{\partial f}{\partial z} - \frac{z}{s}\frac{\partial f}{\partial r}, 0\right) \qquad \text{if} \quad \frac{\partial f}{\partial\varphi} = 0, \end{aligned} \tag{2.98}$$

and, under the assumption of axisymmetry for g,

$$\begin{aligned} \mathbf{q} &= \text{curl}\left[\text{curl}\left(\frac{g}{s}\mathbf{s}\right)\right] = \text{curl}\left(0, \frac{r}{s}\frac{\partial g}{\partial z} - \frac{z}{s}\frac{\partial g}{\partial r}, 0\right), \\ &= \left(\frac{1}{r}\frac{\partial\psi}{\partial z}, 0, -\frac{1}{r}\frac{\partial\psi}{\partial r}\right), \end{aligned} \tag{2.99}$$

with the abbreviation

$$\psi = -r\left(\frac{r}{s}\frac{\partial g}{\partial z} - \frac{z}{s}\frac{\partial g}{\partial r}\right), \qquad s = \sqrt{r^2 + z^2}. \tag{2.100}$$

We thus see: axisymmetric scalar functions f and g define the axisymmetric flow between rotating cylinders if f solves the differential equation

$$v = \frac{r}{s}\frac{\partial f}{\partial z} - \frac{z}{s}\frac{\partial f}{\partial r}, \tag{2.101}$$

where v is the azimuthal component of the flow, and if g solves eq. (2.100), where ψ is the stream function related to the radial and axial components of the flow by eq. (2.108).

We decompose the kinetic energy into its poloidal and toroidal parts. The **dimensionless toroidal kinetic energy** is given by

$$\begin{aligned} E_{tor} &:= c_1 \int_{-\pi/k}^{\pi/k} \int_1^{1+\delta} v^2 r \, dr \, dz \\ &= c_1 \int_{-\pi/k}^{\pi/k} \int_1^{1+\delta} \sum_{\nu=0}^{N} (v_\nu(r) \cos \nu k z)^2 r \, dr \, dz \\ &= c_2 \left(\int_1^{1+\delta} v_o(r)^2 r \, dr + \frac{1}{2} \sum_{\nu=1}^{N} \int_1^{1+\delta} v_\nu(r)^2 r \, dr \right). \end{aligned} \tag{2.102}$$

The integrals are then approximated numerically by the trapezoidal rule

$$\int_1^{1+\delta} v_\nu(r)^2 r \, dr \approx \frac{1}{2} v_\nu(1)^2 \, h + \sum_{\mu=1}^{M} v_\nu(r_\mu)^2 r_\mu h + \frac{1}{2} v_\nu(1+\delta)^2 (1+\delta) h, \tag{2.103}$$

$h = \delta/(M+1)$ being the mesh width and $r_\mu = 1 + \mu h$ being the radius at the μ-th interior grid point. This is an $\mathcal{O}(h^2)$ approximation.

In the same way we approximate the **dimensionless poloidal kinetic energy**

$$E_{pol} := c_1 \int_{-\pi/k}^{\pi/k} \int_1^{1+\delta} (u^2 + w^2) r \, dr \, dz \, . \tag{2.104}$$

The **total kinetic energy** E_{kin} is then given by

$$E_{kin} = E_{tor} + E_{pol}. \tag{2.105}$$

Note that the poloidal kinetic energy of Couette flow is identically zero, and that the dimensionless toroidal kinetic energy of Couette flow does not depend on the Reynolds number, Re.

The **dimensional relative kinetic energies** always depend on Re. They are computed from their dimensionless counterparts in the following way: Using (2.74) and (2.76) for the relation between dimensional and dimensionless quantities we get

$$\begin{aligned} E_{Kin}(\text{Re}) &= c \int_{-\pi/k}^{\pi/k} \int_{1}^{1+\delta} (U^2+V^2+W^2) r \, dr \, dz \\ &= c\,(R_1\Omega_1)^2 \int_{-\pi/k}^{\pi/k} \int_{1}^{1+\delta} (u^2+v^2+w^2) r \, dr \, dz \\ &= (R_1\Omega_1)^2 \, E_{kin}, \\ &= \text{Re}^2 \frac{\nu^2}{R_1^2} \, E_{kin}, \end{aligned} \tag{2.106}$$

and thus

$$\begin{aligned} \frac{E_{Kin}(\text{Re})}{E_{Cr}} &= \frac{(R_1\Omega_1)^2 E_{kin}(\text{Re})}{(R_1\Omega_{cr})^2 E_{cr}}, \\ &= \frac{\text{Re}^2}{\text{Re}_{cr}^2} \frac{E_{kin}}{E_{cr}} \\ &= \frac{T}{T_{cr}} \frac{E_{kin}}{E_{cr}}. \end{aligned} \tag{2.107}$$

Note that the normalization is such that $E_{kin}(\lambda_{cr}, \text{Re}_{cr}) = 1$.

2.4.5 Computation of the stream function

As discussed in the previous subsection, the stream function ψ satisfies

$$u(r,z) = \frac{1}{r}\frac{\partial\psi}{\partial z}, \qquad w(r,z) = -\frac{1}{r}\frac{\partial\psi}{\partial r}. \tag{2.108}$$

Integrating these equations numerically to compute ψ in each grid point would require quite some work. We employed a more efficient, better way. It uses that the values of integrals (2.111) are independent of the path of integration. Let

$$\left(\frac{\partial\psi}{\partial r}, 0, \frac{\partial\psi}{\partial z}\right)^t = \begin{pmatrix} -rw \\ 0 \\ ru \end{pmatrix} =: \mathbf{a}. \tag{2.109}$$

Because of axisymmetry we then get

$$\text{curl}\,\mathbf{a} = \left(\frac{\partial a_r}{\partial z} - \frac{\partial a_z}{\partial r}\right)\mathbf{e}_\varphi = \left(-\frac{\partial(rw)}{\partial z} - \frac{\partial(ru)}{\partial r}\right)\mathbf{e}_\varphi = (-r \,\text{div}\,\mathbf{b})\mathbf{e}_\varphi = \mathbf{0}, \tag{2.110}$$

$\mathbf{b} := (u, 0, w)$ and $\mathbf{e}_\varphi$ the unit vector in angular direction. Now Stokes' theorem shows that integrals (2.111) over closed paths vanish. Thus, given any grid point (r_1, z_1) in our computational domain in the poloidal plane, we compute $\psi(r_1, z_1)$ from the integral

$$\psi(r_1, z_1) = \int_{(r_o, z_o)}^{(r_1, z_1)} r(-w\ dr + u\ dz), \tag{2.111}$$

where starting point (r_o, z_o) and path of integration can be chosen conveniently. We fix (r_o, z_o) to be the lower left corner of the computational domain and choose

$$r(t) = r_o + (r_1 - r_o)t, \quad z(t) = z_o + (z_1 - z_o)t. \tag{2.112}$$

Then $u(r_o, z) = 0 = w(r, z_o)$, and we have to integrate only either $u(r_1, z)$ in the z-direction or $w(r, z_1)$ in the r-direction. Our Fourier decomposition in the z-direction allows us to perform the z-integration analytically, and we get

$$\begin{aligned}\psi(r_1, z_1) &= \int_{(r_o, z_o)}^{(r_1, z_1)} r(-w\ dr + u\ dz) \\ &= r_1 \int_{z_o}^{z_1} \sum_{\nu=1}^{N} u_\nu(r_1) \cos \nu k z\ dz \\ &= r_1 \sum_{\nu=1}^{N} u_\nu(r_1) \frac{\sin \nu k z_1 - \sin \nu k z_o}{\nu k} \\ &= r_1 \sum_{\nu=1}^{N} u_\nu(r_1) \frac{\sin \nu k z_1}{\nu k}\ .\end{aligned} \tag{2.113}$$

The coefficients of ψ at the (ν, μ) interior grid point thus read

$$\psi_{\nu,\mu} = (1 + \mu h) \frac{u_{\nu,\mu}}{\nu k}. \tag{2.114}$$

ψ is given as a sin-series and can be transformed into a function of r and z in the same way as the velocity component w.

2.5 Numerical accuracy

In the sections introducing the methods of discretization used, we explained that the finite differences (2.23)–(2.24) are of order $\mathcal{O}(h^2)$, while Fourier series converge on smooth functions exponentially. These theoretical error estimates are asymptotic results, i.e. they are valid 'for sufficiently small h' (i.e. 'sufficiently large M') and 'sufficiently many Fourier modes' (i.e. 'sufficiently large N'). In this section we report on and discuss investigations of the actual discretization errors experienced with our code.

A question thoroughly investigated in the testing phase was: *Does the code actually show the theoretically expected convergence rates?* The answers obtained in numerical experiments confirmed what we expected, except in one respect: we found a weak dependence of the computed Fourier coefficients on N, the total number of Fourier coefficients. In all cases this dependence was such that it only affected the last digits of the solution in the (r, z)-representation (after the fast Fourier transformations). This dependence is not caused by bugs in the code, but by changes in rounding errors and/or by nonlinear interactions of the discretization errors of the two methods. If the truncation is very serious, such changes can also be caused by nonlinear interactions of modes and aliasing: this was also observed by Marcus [Ma81] in his investigation of how many spherical harmonics are needed in investigations of thermal convection.

A question that we had to ask again and again was the following: *How many grid points, how many modes are adequate?* The equations solved are parameter dependent, and the solutions change their structure when the parameters are varied: the number of vortices changes, surfaces separating different vortices cease to be plane, etc.. Thus the length scale of the spatial structures to be approximated changes with the parameters, and hence the size of the discretization errors. The size of the discretization errors depends not only on the control parameters Re, η and period λ, but also on inner parameters such as the wavelength of a single pair of vortices. Typically it is different on different solution branches for identical values of the control parameters. The number of Fourier modes needed was found to be more strongly influenced by the bifurcation pattern of the solution set than the number of grid points. This is probably related to the fact that the number of vortices inside the computational domain varies more in the axial direction (2 to 16 vortices within a period λ) than in the radial direction (1 or 2 vortices).

Bifurcation points are usually structurally unstable, while fold points are usually structurally stable. Bifurcation points are thus expected to be destroyed by numerical errors, while fold points are expected to persist. This was actually the case for another code with which we compared our results: the finite element code by A. Cliffe [Cli] featured perturbation of bifurcations, especially when the discretization used did not allow the symmetry of the solutions which were to be computed. We do not know of any investigations of the dependence of the existence of fold points on the discretization by Bolstad or Cliffe.

The code Tayperio was found to behave the other way around: bifurcation points turned out to be robust, the existence of fold points sometimes depended on the discretization used. The robustness of the bifurcation points in the code Tayperio is explained by the fact that the coefficients of $\lambda/2$-shifted twins often are identical up to a factor of -1, see (2.81) and the text following it. This enforcement of symmetry preserves bifurcations.

We now report on the details of the investigations. Those investigations mostly performed with a fixed number of modes and several different numbers of radial grid points are reported in the subsection 'Finite Differences', while those investigations which focused on the number of modes needed are reported in the subsection 'Truncation of the Fourier Series'. The remarks on the interaction of both types of truncation errors are also in this subsection.

2.5.1 Finite Differences

Computations with radial mesh widths $h = 1/8, 1/16, 1/32$, and $1/64$ were carried out with different numbers of Fourier modes. When fixing the number of modes and varying the number of grid points we did observe quantitative, but never qualitative changes. The computations leading to Fig. 2.5 and to Fig. 2.6 both show very clearly that the theoretically expected $\mathcal{O}(h^2)$ error dependence is obtained and that the value $h = 1/32$ used in many of our computations yields good approximations for moderate Reynolds numbers. Fig. 2.5 repeats Fig. 1.3, but in addition it shows computations with fewer grid points. It thus allows us to compare computations with $h = 1/8,\ 1/16,\ 1/32$ and as many Fourier modes as were required such that the discretization error of the Fourier decomposition was negligible (i.e. the highest two Fourier components had the same size as the rounding errors). The quadratic convergence in h can clearly be seen, and $h = 1/32$ gives quite acceptable values $\left(\frac{G(R_2)-G(R_1)}{G(R_1)} < 0.5\%\right)$. Note that even the

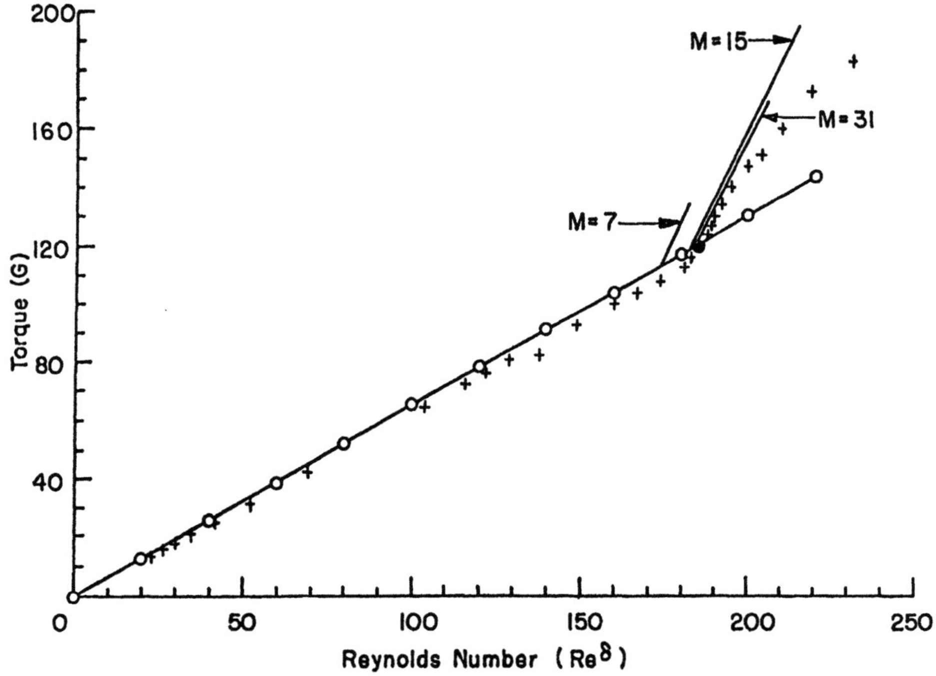

Figure 2.5: Torque vs. Re$^\delta$ for narrow gap case: $\eta = 0.95$, $\mu = 0$, $\lambda = 2.007$. $+$: experiments of [DS]; $- \circ -$: Couette flow; —: computations with Tayperio; $\bullet$: accurate computed bifurcation point, $M = 80$. Three computations of the bifurcating branch are shown, each using a different number M of interior grid points and $N \leq 6$ modes [MK80, Fig. 1].

computations with $h = 1/8$ give results which are *qualitatively* correct: the curve computed with $h = 1/8$ has the correct inclination, it is only displaced in the parameter space. For grids which are too coarse, bifurcation occurs at a Reynolds number which is too small. This results a shift of the marginal curve; see the investigations for $\eta = 0.5$ and various values of h in Fig. 2.6.

Figure 2.7 shows outward transition solutions which were computed with enough Fourier modes, $N = 6$, but with too few grid points, $h = 1/8$. Nevertheless, the structure of the solutions is essentially correct.

Figure 2.8a shows a solution whose level lines suffer from discretization errors both of the Fourier decomposition and of the difference method. Note the high frequency wiggles introduced by the plotting procedure (linear interpolation). They are indicative of large gradients and thus suggest

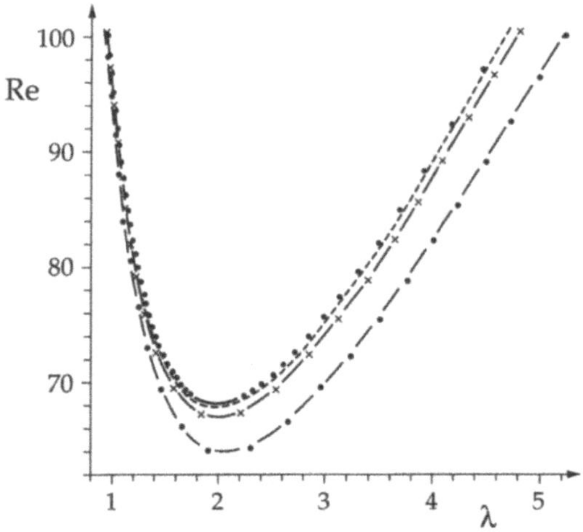

Figure 2.6: Five approximations to the neutral curve for $\eta = 0.5$, $\mu = 0$. Three are computations with the code Tayperio, using M interior grid points. $M = 7$: —•—; $M = 15$: —×—; and $M = 31$: – – –. The innermost curve ··· was computed by Richardson extrapolation from those for $M = 15$ and $M = 31$. At its minimum, it agrees well with the values given in Table 1 of [DPE] —, obtained from expansions up to 5th order [FM, Fig. 1].

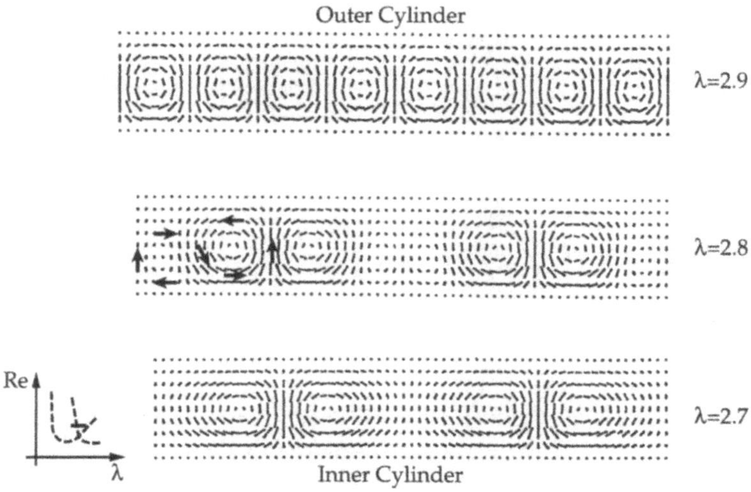

Figure 2.7: Inward transition solutions, computed with enough Fourier modes, $N = 6$, but with too few grid points, $M = 7$; ($\eta = 0.5$, $\mu = 0$) [FM, Fig. 3].

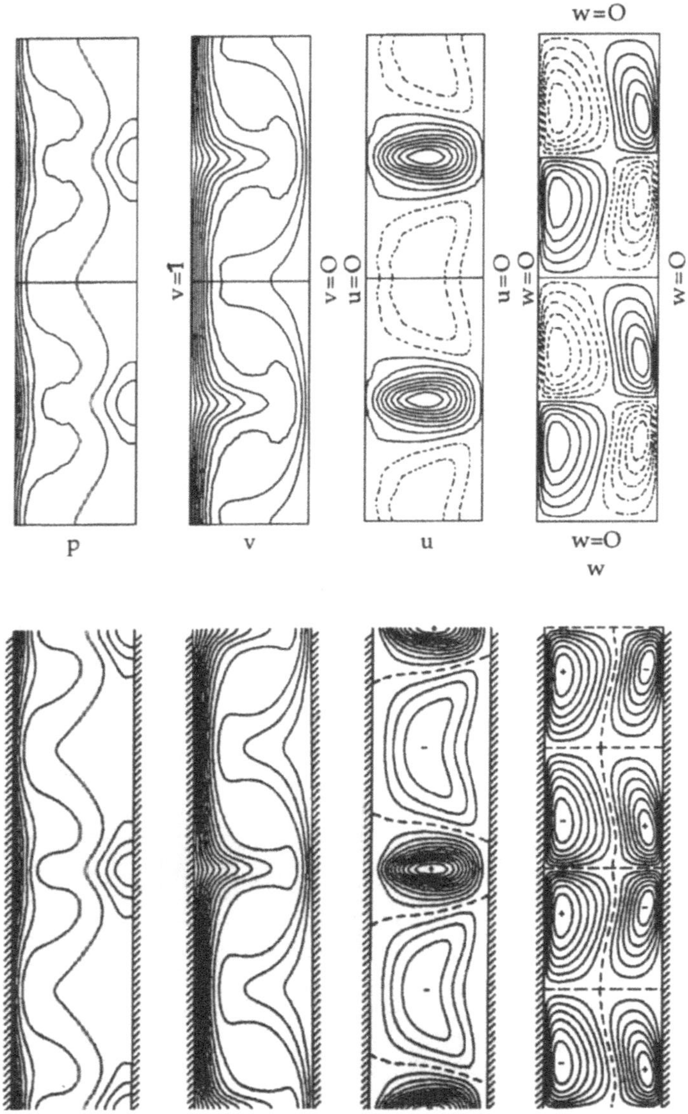

Figure 2.8: Comparison of solutions in different discretizations (the solutions happen to be shifted by $\lambda/2$ with respect to each other).
a) Solution for Re $=$ 4.19 Re_{cr}, $\eta = 0.5$, $\lambda = 1.988$ as computed with $N = 8$, $M = 15$ [MK80, Fig.4].
b) Solution for Re $=$ 4.77 Re_{cr}, $\eta = 0.5$, $\lambda = 2$ as computed by Booz with $\mathcal{O}(h^4)$–finite differences, 73×121 grid points, and fully implicit in time; taken from [Boo, Fig.26].

the need for finer meshes. Also the number of Fourier modes was not sufficient as all computed coefficients were of comparable orders showing that some absent modes were non-negligible. Nevertheless, the structure of the approximate solutions is essentially correct, as we can see from comparison with Fig. 2.8b. The Reynolds numbers of cases a) and b) are not identical, but the structure of the solutions did not change much when we computed along the branch with continuation in Re (compare the flow for $\mathrm{Re} = 1.9\,\mathrm{Re}_{cr}$ shown in Fig. 3.1b). We observe again that the discretization error in the finite differences shifts structures to lower Reynolds numbers, as in Figs. 2.5 and 2.6.

2.5.2 Truncation of the Fourier Series

The errors caused by truncating the Fourier expansions (Galerkin method) are of a different nature than those in finite difference approximations. As long as only a few harmonics are needed, this is easily detected by the fact that the coefficients of the higher harmonics are practically zero. For many of our computations [MK80, FM, MK85] we could thus use as many harmonics as needed. $N \leq 6$ was sufficient to keep the last two or three coefficients small. In later computations [MW1, MW2, SWM] we encountered branches which required more harmonics than the available storage on the Cray 1 or Cray X-MP allowed. No theoretical criterion seems to be known how to balance the errors of the two methods. So we repeatedly did numerical experiments for finding adequate discretizations.

What happens when solutions are approximated with too few harmonics?

When the same solution was computed with N_1 and $N_2 > N_1$ Fourier modes, we often observed that the smaller ones of the N_1 Fourier coefficients changed a bit, but only in the last decimals[1]. An N-dependence of the first digits was observed only in coefficients that were small enough to be neglected. When N_1 was 'much too small', (for instance $N_1 = 2,\ N_2 = 6$, and 4 coefficients of the N_2-representation clearly above rounding level), also the first coefficients could change a bit. Whenever the available computer storage allowed, $N_2 - N_1 = 2$, and the change was made 'before it became necessary'.

Truncation errors easily cause *qualitative* changes. This was investigated by several authors phenomenologically in numerical experiments, especially in connection with the Lorenz equations [Ma81], [GH, p.92]. In our

[1] 6 digit output

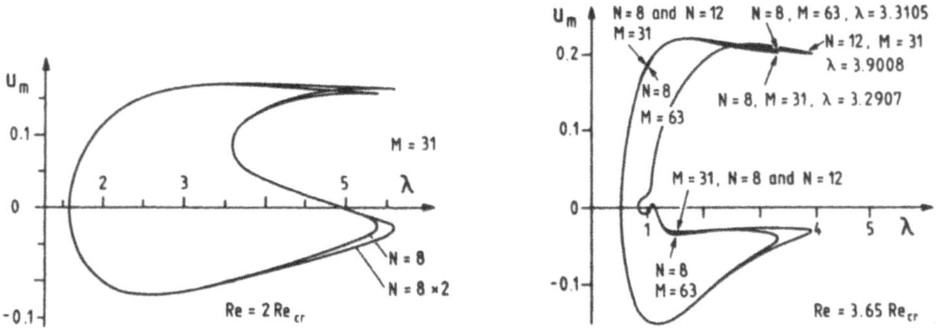

Figure 2.9: Comparison of computations with different discretizations [SWM, Fig. 13].
a) $\times 2$-branch, Re $= 2\text{Re}_{cr}$, computed directly with $N = 8$ and the λ's shown, and computed as periodic repetition of the $\times 1$-branch, using $N = 8$ for $\lambda/2$. This corresponds to computations with $N = 16$, forcing every second Fourier coefficient to be identically zero. (Discretisation errors for $N = 8 \times 2$-computations are the same as for $N = 16$, rounding errors are different.)
b) u_m versus λ plot of the basic 2-vortex branch for Re $= 800 = 3.65\text{Re}_{cr}$ as computed with $M \leq 63,\ N \leq 12$.

case, it is strongly felt when we continue in λ, and it is easy to understand theoretically. For example, 1 harmonic is sufficient to approximate the 2-vortex solutions that bifurcate from Couette flow for Re $= 2\text{Re}_{cr}$ and $\lambda \approx 0.78$. We need 2 harmonics to approximate the periodic repetition of this solution on the $\times 2$-branch with the same numerical accuracy. And we need 8 harmonics to approximate it on the $\times 8$-branch. For approximating mixed-mode solutions appropriately, the modes interacting have to be present, see Fig. 2.11 and its discussion.

In some of our computations we defined

$$\varepsilon := \frac{a_N}{a} \tag{2.115}$$

where a_N is the maximum absolute value of the highest Fourier coefficient and a is the maximum absolute value of all Fourier coefficients for that solution. We required at least $\varepsilon < 10^{-1}$. For Fig. 3.33, for instance, we had $u_6(1 + \delta/2)$ as maximum coefficient, and computations with $N = 14$ generated only $\varepsilon < 10^{-1}$.

Though it is unfortunate that solutions can disappear, the periodicity present in the structure of the solution set gives us a convenient tool to es-

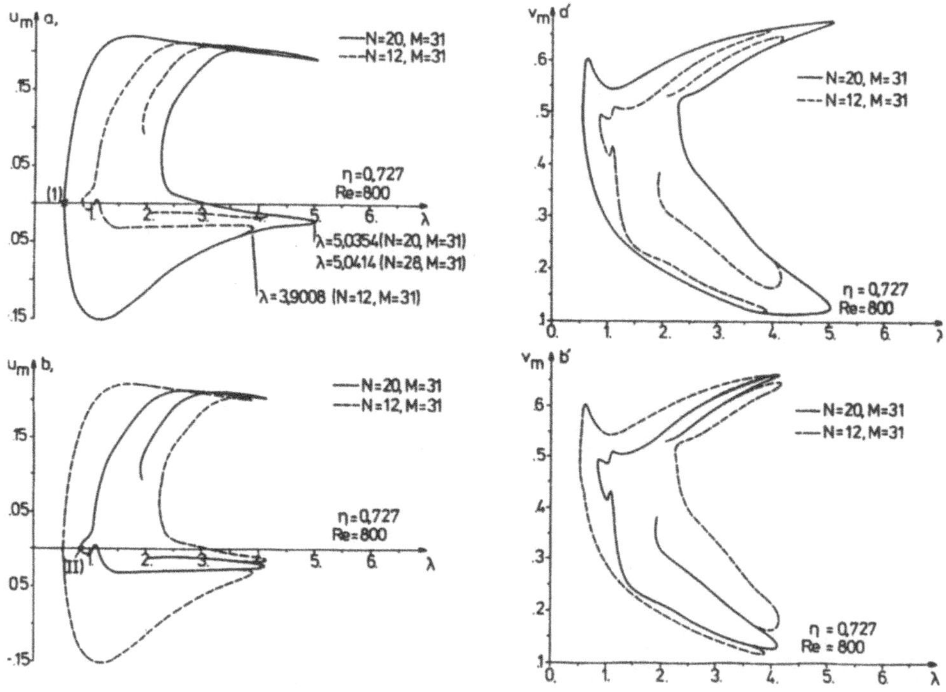

Figure 2.10: Change of the connectivity of neighboring solution branches with discretization. a), a'): Computed with $N = 12$, the basic 2-vortex branch bifurcating at (1) from Couette flow changes continuously into the basic double vortex branch and bifurcates at (II) to Couette flow. Computed with $N = 20$ or $N = 28$, the branch bifurcating at (1) from Couette flow meets its twin at a wavelength-halving bifurcation to the $\times 2$-branch of the double-vortex flow, i.e. at Point 1 of Fig. 3.11. Computed with $N = 12$ in the reversed direction, the branch starting at Point 1 passes Points 5 and 8 and goes on as displayed in Fig. 3.11. b), b'): Same type of investigation, starting at Point (II).

The deviations between the computations for $N = 28$ and for $N = 12$ occur only in a neighborhood of the fold points, for $\lambda \geq 3.5$. Whenever a dashed line coincides with a solid line, only a solid line is seen. In the $N = 28$ Fourier representation of $u_m = u(1+\delta/2, 0)$ at the tip of one of the folds ($\lambda = 5.0414$), the first 8 harmonics are of order 10^{-2}, the next 9 of order 10^{-3}, the next 8 of order 10^{-4}, and only the last 3 are of order 10^{-5}. Thus $\varepsilon < 10^{-2}$ [SWM, Fig. 14].

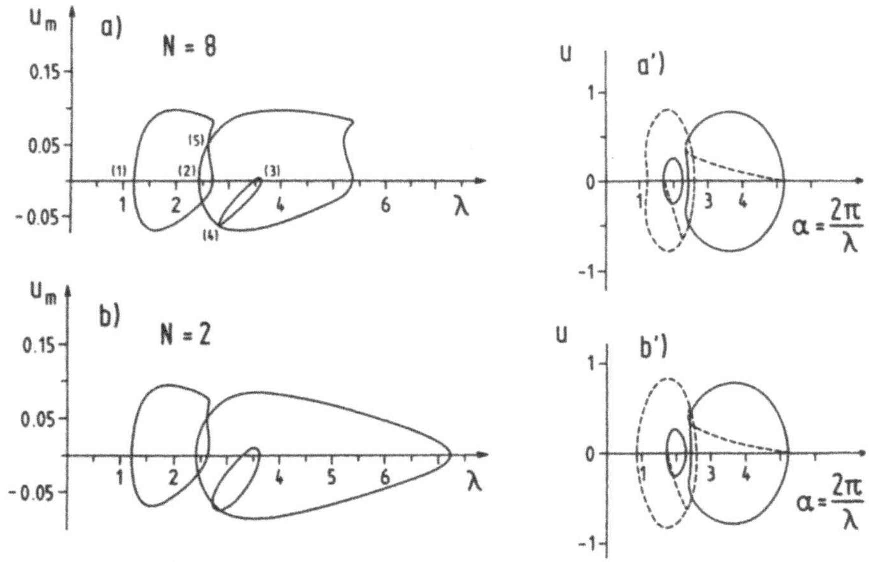

Figure 2.11: Discretization error for $Re = \sqrt{1.5}Re_{cr}$, $\eta = .727$, $\mu = 0$.
a) Radial velocity u_m versus λ as computed with practically no Fourier discretization error ($N = 8$ harmonics).
a') First (—) and second (- - -) harmonic at the midpoint (r_m, z_m), $r_m = 1 + \delta/2$, $z_m = 0$, versus wave number α. *b*, *b'*) : Same branches as in *a*, *a'*), but with strong Fourier discretization error ($N = 2$). Details are given in the text [SWM, Fig. 15].

timate the size of the actual error. This is demonstrated in Fig. 2.9a, where we display the result of two sets of computations which were performed with N = 8 Fourier harmonics each. We see that N = 8 is sufficient for $\lambda \leq 3$, while more harmonics should be used when approaching the fold points.

The fact that the loci of fold points are more sensitive to truncation of the Fourier series than to finite-difference errors is demonstrated in Fig. 2.9b. The main reason for this sensitivity seems to be that the folds of Fig. 2.9 are generated by nonlinear interactions of the branches shown with other branches which require more harmonics for their representation. In the case of Fig. 2.9a, it is the interaction of the computed $\times 2$-branch with the $\times 4$-branch not shown in the figure. Note that an $N = 8$ approximation of the $\times 4$-branch has only 2 coefficients which are nonzero above the level of rounding errors – the 4th and the 8th.

After the computations for Fig. 2.9b were finished, the CRAY 1 of ZIB Berlin was replaced by a CRAY X-MP. This larger machine allowed computations with $h = 1/32$ and $N \leq 28$. Repeating the computations which led to Figs. 3.9a, 3.11 and 3.21 showed that the separate but neighboring branches displayed in Figs. 3.9a and 3.11 are connected when recomputed with more Fourier modes. The need for recomputing was felt because the branch of Fig. 3.11 had been found by chance, by using too large steps in pseudo-arclength near the fold of the branch of Fig. 3.9a. That those steps were too large was detected because we wondered about the results obtained, recomputed with smaller step size and got different results. In the computations leading to Fig. 2.10, we used very small steps, to make sure that the results will not change when recomputed with smaller steps in the pseudo-arclength.

Computed with $N = 12$, the basic 2-vortex branch bifurcating from Couette flow at Point (1) of Fig. 2.10 changes continuously into the basic double vortex branch and bifurcates at (II) to Couette flow (as in Fig. 3.9a). Computed with $N = 20$ or $N = 28$, the branch bifurcating at (1) from Couette flow ends at a wavelength-halving bifurcation to the $\times 2$-branch of the double-vortex flow, i.e. it ends (together with its twin) at Point 1 of Fig. 3.11. Computed with $N = 12$ in the reversed direction, the branch starting at Point 1 passes Points 5 and 8 and then goes on as displayed in Fig. 3.11[2]. Similar computations starting at Point (II) are shown in the lower part of Fig. 2.10. They show the form of the other branch for $N \geq 20$ and confirm that all 4 surfaces meeting in the two folds numerically exist under all discretizations we used. The deviations between the computations for $N = 28$ and for $N = 12$ occur only in a neighborhood of the fold points, for $\lambda \geq 3.5$. This was to be expected from the analysis of Fig. 2.9b.

Usually, fold points are structurally stable. Figure 2.10 shows that results on the existence of folds can be falsified by discretization errors. It should be noted, however, that these investigations were done in a very special parameter regime. This will be further discussed in Chapter 3, near Fig. 3.20.

Nevertheless, what was written about this in our discussion of the basic (2, 4)-fold [MW2] is correct: "The form of a fold, its locus in parameter space, and even its existence for higher Reynolds numbers are subject to discretization errors. The existence of the two surfaces meeting in the fold,

[2] In Fig. 2.10, the display of portions of the branch shown in Fig. 3.11 always ends at the fold point 8.

however, and also the structure of the solutions on these surfaces, do not seem to be affected by discretization 'far away' from the fold."

This 'far away' is made more precise in the following example which is 'completely under control'. $N = 8$ gives computations 'without Fourier discretization error' for the branches displayed in Fig. 2.11a, a$'$. $N = 2$, however, is a severe truncation: the $\times 4$-branch interacting with the $\times 2$-branch does not exist in this approximation. As is shown in 'Computations for Re $= 2$ Re$_{cr}$' in the next chapter, the basic $(4, 8)$-fold and the secondary bifurcations $(5)\times 2$ and $(4)\times 2$ are generated by the $(4, 8)$ double eigenvalue, which does not exist for $N = 2$ either. It is thus no surprise that this fold and the secondary bifurcations completely disappear for computations with $N = 2$. This change of structure, however, could easily be detected by computations with $\lambda/2$ instead of λ.

In the case of the closed solution curves of Fig. 3.26, such a test was not available since those branches have minimal period. It might thus turn out that the form of the curves (i.e. not only the loci of the folds in parameter space but also the way in which the branches are joined together) is affected by discretization errors. Nevertheless we think that these computations give some ideas about the structure of the solution set.

2.5.3 Conclusions

We have seen that the size of the discretization error is strongly dependent on which of the (possibly several) solutions is approximated for which parameter value. This especially applies in the neighborhood of fold points. For Taylor vortex flows, many fold points arise from perturbed bifurcations. Thus the structure of the flows often changes near fold points. From this it follows that it is not adequate to compute a solution manifold with the same number of grid points/modes in a large parameter domain. It especially does not make sense to estimate the discretization errors just *once* for a specific solution/parameter value and then stick to the number of grid points/modes necessary there for all later computations. The size of the discretization errors has to be investigated *again and again* since it depends so much on the parameter values, and on which solution branch is computed. This should be kept in mind when using fold following or bifurcation point following algorithms: the tests on the size of the discretization errors should be made for those parameter values where they are expected to be largest.

Chapter 3

Stationary Taylor Vortex Flows

In this chapter we discuss stationary Taylor vortex flows, focusing on results that were obtained by using the code Tayperio. The numerical approach of the code was discussed in Chap. 2, and also the validity of the underlying numerical model. Some remarks on methods employed by other authors can be found here, and also in Chapters 1 and 4.

3.1 Introduction

In this chapter we study stationary axisymmetric flows in a wide gap between concentric cylinders as described by the Navier-Stokes equations in eqs. (2.1)–(2.3). End effects are neglected (periodic boundary conditions). With both cylinders rotating, there are four parameters in the problem: the Reynolds number Re, the axial period λ, the radius ratio η, and the rotation ratio μ.

In the following we shall discuss questions such as: What types of flows are allowed by eqs. (2.1)–(2.3) in the highly nonlinear regime ($\mathrm{Re} \gg \mathrm{Re}_{cr}$)? How many solutions exist for a given set of parameter values? What is the structure of the solution branches as Re and λ vary? What folds and bifurcations do they undergo?

At present, to prove theorems answering these questions for the parameter values used in our computations seems to be out of range [Lad69], [Te83, §§9, 10]. Our "experimental" investigation of the solution set of eqs. (2.1)–(2.3) is thus a contribution towards a better understanding of the mathematical properties of the Navier-Stokes equations. An example of how such investigations can help to formulate realistic conjectures about these equations is given in this chapter and pursued in the next chapter: it concerns the global structure of secondary bifurcations. To our knowledge, it was the first result of this kind when published by Meyer-Spasche et al. in [MW2, LMP].

The mathematically motivated investigations of the solution set near the secondary bifurcations turned out to be of greater interest to physics than was expected during the course of these investigations: the stability of the stationary vortices in wide gaps is governed by the Eckhaus and the short-wavelength instabilities. The neutral curves for both instabilities are strongly influenced by the secondary bifurcations. This is illustrated by Figs. 3.37 and 3.40.

We solve the Navier-Stokes equations (2.1)–(2.3) numerically, using the very reliable methods described in Chap. 2: Discretization by Fourier decomposition in the axial direction and centered finite differences in the radial direction; systematic variation of Re and λ by using the method of continuation with Newton-Gauss iterations. For most of our investigations we fixed η and μ at $\eta = 0.727$, $\mu = 0$, which was shown to allow axisymmetric steady flows up to Re > 9 Re_{cr} for a large interval of possible λ's [Jo81, BuKo].

In a first set of computations, bifurcation from Couette flow to stationary vortices was accurately determined, and the bifurcating flows were computed for fixed period $\lambda_{cr} \approx 2$ and increasing Reynolds number Re. This was the first application of the pseudo-arclength continuation methods of Keller [Ke77] to partial differential equations. Agreement with the experiments of Donnelly, Simon was extremely good for the wide gap case ($\eta = 0.5$), and solutions were obtained for a wider range of Reynolds numbers than previously reported.

In the following computations, questions of existence and of properties of flows for periods $\lambda \neq 2$ were pursued. Solutions consisting of vortices of different sizes were detected. Varying both the period λ and the Reynolds number, it was found that these solutions are intermediate states between solutions with equally sized vortices (transitions from 2 to 4 and from 2 to 6 vortices). Most of these computations were performed for $\eta = 0.5$. Subcritical bifurcations from Couette flow were found to occur for certain parameter ranges.

In the course of these computations it was found that pseudo-arclength continuation with respect to the period λ was necessary for understanding the bifurcation patterns encountered. After that was implemented, the variation of the flows with changing period and Reynolds number could be investigated more systematically. Closed loops of solution branches and secondary bifurcations were found. A λ-interval within the neutral curve for (Couette flow, 2-vortex flow) was found such that no 2-vortex solutions could be computed for these λ-values.

Later investigations showed that this non-existence is actually due to the forced broken symmetry of the Taylor apparatus: it causes nonlinear interactions of the solution branches. Some secondary bifurcations existing in a more symmetric configuration are perturbed, and thus some of the solution branches close in a different way than they would in the more symmetric configuration, i.e. in the case of the Rayleigh-Bénard problem. A comparison of investigations by Busse for the perturbed Rayleigh-Bénard problem with wide gap computations clarified the situation (Fig. 3.18). One of the dominant features of these closed solution branches is the basic $(2,4)$-fold and its periodic repetitions, the basic $(n,2n)$ folds. In later computations this fold was found to exist up to at least $\mathrm{Re} = 4.6\ \mathrm{Re}_{cr}$.

The curve of neutral stability of Couette flow in the (Re,λ)-plane can be found by solving a 'linear' eigenvalue problem (first eigenvalue, Taylor 1923). It was found that flows bifurcating from Couette flow at higher eigenvalues interact in the way just described with the basic flow bifurcating at the first eigenvalue. They thus play an important role in the analysis of the basic flow. The two other most important flows for moderate Reynolds numbers are those bifurcating at the second eigenvalue in the axial direction ('4-vortex flow', 'periodic repetition') and in the radial direction ('double-vortex flow'). Double-vortex flows consist of two adjacent vortices in the radial direction, which gives rise to free stagnation points of the poloidal flow. It was found that the secondary bifurcations and the $(2,4)$-fold are generated by the $(2,4)$ double point $(\mathrm{Re}_{24},\lambda_{24})$, the intersection point of the (Couette flow, 2-vortex flow) neutral curve with the (Couette flow, 4-vortex flow) neutral curve. Further investigations gave evidence that one of the curves of secondary bifurcations also emanates from the point $(\mathrm{Re}_{d4},\lambda_{d4})$, the intersection point of the (Couette flow, 4-vortex flow) neutral curve with the (Couette flow, double-vortex flow) neutral curve. This suggests that this curve of secondary bifurcations connects the two double points $(\mathrm{Re}_{24},\lambda_{24})$ and $(\mathrm{Re}_{d4},\lambda_{d4})$, cf. Fig. 3.24.

These findings are further pursued in the next chapter:

On the one hand, three simplified model problems derived from the Rayleigh-Bénard problem are analyzed. For these problems no bifurcations are perturbed, and the loci of the primary and secondary bifurcation points can be explicitly computed in the (Rayleigh number, wave number) plane. In the further simplified case of infinitely large Prandtl number it is shown analytically that a line of secondary bifurcations does connect the intersection points of neutral curves corresponding to the points $(\text{Re}_{24}, \lambda_{24})$ and $(\text{Re}_{d4}, \lambda_{d4})$. For other Prandtl numbers it was confirmed numerically.

On the other hand we report on computations by Bolstad, who followed several of these singularities with a fold following and bifurcation point following algorithm in the (Re, λ)-plane and thus confirmed the conjectures.

A large multiplicity of solutions (≥ 21 for $\lambda = 2.5$, not counting $\lambda/2$-shifted twins) and of fold points (≥ 25) and period doubling and tripling bifurcations were detected for $\text{Re} = 3.65\ \text{Re}_{cr}$, $\eta = 0.727$, $\mu = 0$, λ varying. Solutions with free stagnation points of the poloidal flow and with other unusual flow patterns were found and are displayed in this text.

When Re is fixed at the more moderate value $\text{Re} = 2\ \text{Re}_{cr}$ and λ varies, nested families of 'isolas' (closed loops of solution branches) are found. We discuss how a single isola is formed by interaction of secondary branches belonging to 3 different intersection points of neutral curves. New flow patterns were found. Quite a number of them have a separatrix.

In the course of these investigations, it was found that increasing Reynolds number and increasing influence of the nonlinear terms of the equations required finer discretizations. It was thus considered necessary to check the numerical accuracy of the computations again and again, and to adapt the number of grid points and Fourier modes as needed. A new result of the investigations reported here seems to be that the bifurcation structure of the solution set strongly influences the resolution required for the computations. The quality of the numerical approximations is discussed in Chap. 2, since the details depend on the numerical properties of the code used. The range of Re-numbers accessible in this approach is limited by the available computer storage, not by the method.

At the end of the present chapter we discuss the stability of stationary Taylor vortex flows, reporting results by Jones and Riecke, Paap. As mentioned earlier, the secondary bifurcations and fold points investigated play an important role in the stability of Taylor vortex flows to axisymmetric perturbations.

3.2 Computations with fixed period $\lambda \approx 2$

For the purposes of debugging the new program, of testing the validity of the numerical model and the power of the numerical methods used, extensive computations were done on two configurations for which experimental results [DS] were available. In these experiments of Donnelly/Simon, the outer cylinder was always at rest, $\mu = 0$; the outer radius was $R_2 = 2$ cm; two values for the inner radius R_1 were used: $R_1 = 1$ cm, $D = 1$ cm, the *wide gap* case, and $R_1 = 1.9$ cm, $D = 0.1$ cm, the *narrow gap* case. The experimental cylinders were 10 cm long and the torque was measured on the middle 5 cm of the outer cylinder, to reduce end effects (see the discussion in Chap. 1). By using different fluids and by varying the inner angular velocity Ω_1, wide ranges of Reynolds numbers were covered in both cases.

In our code, the torque of Couette flow and of the bifurcating stationary vortices were computed according to formulas (2.87) and (2.89), as described in the previous chapter. The computed values were then compared to the measured values of Donnelly/Simon [DS]. For $\eta = 0.5$ agreement between measurements and computations was excellent in the full Re-interval that was accessible with acceptable numerical error. In the case $\eta = 0.95$ agreement was not quite that good, both for subcritical and supercritical Reynolds numbers. The torque versus Re^{δ} plots are shown in Figs. 1.2 and 1.3 and discussed below.

Before the computations could start, a decision had to be made about the wavelength of the vortices to be computed:

- in the experiments, the wavelength λ is an *inner parameter (behavior parameter)*, which is not at the disposal of the experimenter. The flow 'chooses its own wavelength' in dependence on the previous history of the flow, i.e. in dependence on the initial conditions. It thus has to be measured in experiments, and that was not as easy in 1960 as it is today. Donnelly/Simon didn't give the wavelengths of the flows leading to the measured torque.
- in our numerical model, the period λ is an *outer parameter (control parameter)* and has to be fixed before we start computing.
  This discrepancy between experiments and computations is a consequence of the periodic boundary conditions.

Donnelly, Simon increased the Reynolds number quasistationarily. We assumed that in this case Taylor vortices would first appear at Re_{cr} with wavelength λ_{cr}, i.e. at those values which correspond to the minimum of the marginal curve in the (Re, λ)-plane, see Figs. 1.9 or 2.6. (This is supported by the experiments by Burkhalter/Koschmieder, see the section 'end effects' in Chapter 1). Those first computations were thus performed with period λ_{cr}.

3.2.1 A narrow gap problem, $\eta = 0.95$

The fluid used in these experiments had $\nu = 5.796 \times 10^{-3}$ cm^2/sec and $\rho = 1.585$ g/cm^3. Computations were done for a series of Re values while the axial wavelength λ was fixed at $\lambda_{cr} = 2.007$, the value at the minimum of the marginal curve as computed by DiPrima/Eagles [DPE][1]. Figure 1.3 is a torque versus Re^δ plot, Re^δ as defined in (1.10). We give both the torque of the z-independent Couette solution (1.12) computed from formula (2.87), and the average torque of the bifurcated solution, computed from formula (2.89) as described in Chapter 2, and evaluated at R_1 with $L_o = 5$ cm. The curve of the bifurcated branch was drawn by linear interpolation between 22 computed points: 12 values in $184.21 \le \mathrm{Re}^\delta \le 188.66$ with $N = 2$ Fourier modes, and 10 values in $189.28 \le \mathrm{Re}^\delta \le 205.35$ with $N = 4$ Fourier modes. The number of radial interior grid points was $M = 31$.

An extensive analysis of the discretization error was performed in connection with these computations. It is discussed in Chapter 2. Figure 2.5 is a different version of Fig. 1.3 and shows how the computed values change with refinement of discretization.

There are deviations between the computed and measured values of the torque that are not due to numerical error. The deviation for $\mathrm{Re} < \mathrm{Re}_{cr}$ was already discussed in Chapter 1. The deviation for $\mathrm{Re}^\delta > 195$ is also not due to numerical error, but to the emergence of *wavy vortices*. The question if this deviation is due to the fact that the flow becomes 3-dimensional and periodic in time at these parameter values was investigated by Ken Meyer [Mey67, Fig. 5], [Mey69] with one of the first codes for Taylor vortex flows. It was theoretically confirmed in [DDS, Eg71, Eg74] and by Jones [Jo81]. This is further discussed in the section on stability of Taylor vortex flows at the end of this chapter.

Figure 3.1a shows level lines of the reduced pressure p, of the velocity components u, v and w, and of the (u, w) vector field, over the (r, z)-plane for $\lambda = 2.007$ and $\mathrm{Re}^\delta = 184.3 \approx \mathrm{Re}^\delta_{cr}$.

[1] In [DPE] α values are given, $\alpha = 2\pi/\lambda$.

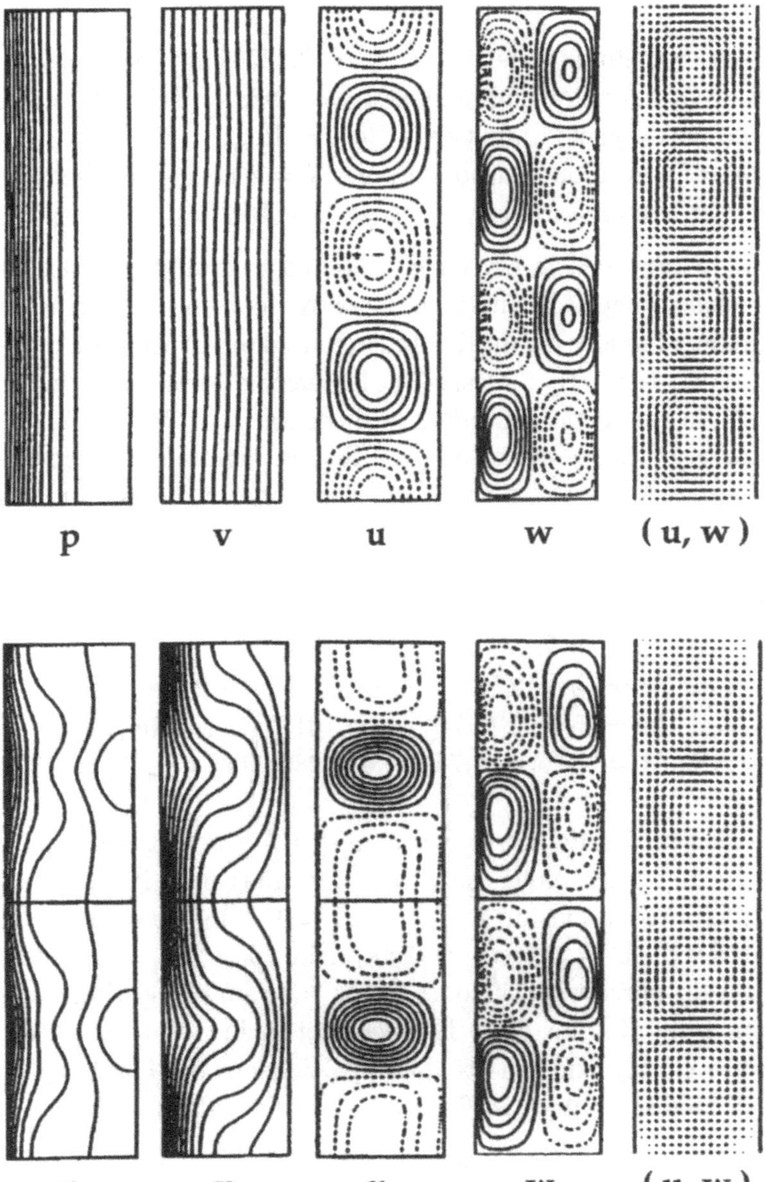

Figure 3.1: Level lines of the reduced pressure p, of the velocity components u, v and w, and of the (u, w) vector field over the (r, z)-plane for $\mu = 0$.
a) $\eta = 0.95$, $\lambda = 2.007$, $\mathrm{Re} \approx \mathrm{Re}_{cr}$, with N=2, M=31 [MK78, Fig. 4.2].
b) $\eta = 0.5$, $\lambda = 1.988$, $\mathrm{Re} = 1.9\ \mathrm{Re}_{cr}$, with N=8, M=15 [MK78, Fig. 4.13].

3.2.2 A wide gap problem, $\eta = 0.5$

In this case the fluid properties were $\nu = 0.1226$ cm^2/sec, $\rho = 0.8404$ g/cm^3, and $\mathrm{Re}_{bif}(\lambda)$ has its minimum at $\lambda = 1.988$ [DPE]. The corresponding value of the Reynolds number is $\mathrm{Re}_{cr} = \mathrm{Re}^{\delta}_{cr} = 68.113$ when computed with $M = 63$ interior grid points (and $N = 2$ Fourier modes[2]). Flows stay axisymmetric and stationary up to at least $\mathrm{Re}/\mathrm{Re}_{cr} > 4$ [Jo81, DS].

We found excellent agreement between measured and computed values in the whole range of Reynold numbers treated. But it became evident that the number of modes and grid points required for an accurate representation of the solutions greatly increased for increasing Re values. Using the criteria discussed in the section on 'numerical accuracy' in Chapter 2, we increased the number of modes up to $\mathrm{Re} \leq 1.5\ \mathrm{Re}_{cr}$ as required, but kept the number of grid points fixed after a first short study on the necessary number of grid points. The bifurcating branch was computed with

$$N = 4,\ M = 31 \text{ for } \mathrm{Re} \leq 73;$$
$$N = 2,\ M = 15 \text{ for } 67.05 \leq \mathrm{Re} \leq 67.25;$$
$$N = 4,\ M = 15 \text{ for } 67.25 \leq \mathrm{Re} \leq 81.17 \approx 1.2\ \mathrm{Re}_{cr};$$
$$N = 6,\ M = 15 \text{ for } 81.17 \leq \mathrm{Re} \leq 101.76 \approx 1.5\ \mathrm{Re}_{cr};$$
$$N = 8,\ M = 15 \text{ for } 101.76 \leq \mathrm{Re} \leq 280.99 \approx 4.13\ \mathrm{Re}_{cr}.$$

Some of these results are shown in Fig. 1.2, where the computed torque versus Re is plotted along with the values measured in the experiment of [DS]. Computed values up to $\mathrm{Re} \leq 127 \approx 1.9\ \mathrm{Re}_{cr}$ are shown. The values computed with $M = 31$ were also plotted in this figure. They cannot be seen because they agree with the ones for $M = 15$ up to plotting accuracy. The solutions computed for $\mathrm{Re} > 127$ were not included in this figure, because they were clearly affected by discretization errors. The torque values of these omitted six computed flows are given in [MK78, p. 52].

In Fig. 2.8 contour plots of p, u, v, w are shown for $\mathrm{Re} = 4.13\ \mathrm{Re}_{cr}$. Comparison with computations by Booz [Boo] (solution of the time-dependent equations with 4th order accuracy) shows that this solution is still qualitatively correct though the discretization errors were large. For more details see the section on numerical accuracy, Chapter 2. The corresponding plots for $\mathrm{Re} = 126.9 \approx 1.9\ \mathrm{Re}_{cr}$ were still quite smooth, see Fig. 3.1b.

[2]Only the first Fourier mode contributes.

3.3 Variation of flows with period λ

3.3.1 Previous results on flows with wavelengths $\lambda \neq 2$

As was illustrated and discussed in Chap. 1, there are large parameter regimes in which the control parameters do not uniquely determine the flow. And the number of observed flows varies widely with the control parameters. The three different flows shown in Fig. 1.8, for instance, were obtained for $\eta = 0.727$, $\mathrm{Re} \approx 3\ \mathrm{Re}_{cr}$. The inner parameter 'wavelength λ' varies from $\lambda = 1.530$ to $\lambda = 2.405$. This is quite a wide range, but not the full range of λ values situated inside the marginal curve of Couette flow, i.e. it is not the full range that had to be considered theoretically possible when we started the computations to be discussed now. By experiments alone, the following questions could not be answered in principle: Didn't they find other flows because they are unstable, because they don't exist, or because the experimenters didn't succeed in finding the necessary experimental conditions, i.e. the proper initial values? In the case of stationary flows, later research showed that each of the three explanations is correct, depending on the parameter regime considered. This research was done in interaction of calculations (using perturbation methods), computations (using numerical methods), and experiments.

Existence proofs

A first existence proof for Taylor vortex flows was given by Velte [Ve66], in connection with a bifurcation analysis at the neutral curve of Couette flow. He proved existence above the neutral curve in the (λ, Re)-plane, except at intersection points of the curve with its own periodic repetitions. Later existence proofs showed that the number of existing solutions is always odd [Te83]. Thus solutions always appear or disappear in pairs when the parameters are changed.

The parameter values at which such changes happen (because of bifurcations, folds etc.), and thus the number of existing solutions at given parameter values, cannot be obtained with analytical methods alone far away from analytically known solutions. In these parameter regimes, results on numbers of solutions can be obtained by experiments and computations only. This is no rigorous proof of their existence as solutions of the underlying mathematical model. In the computations, we can investigate the convergence properties of the discretization used (see Chap. 2), but it is practically impossible to verify in detail the assumptions of the convergence theorems [Me75]. Also, we can never be sure that we found all solutions when doing computations/experiments. When using pseudo-arclength continuation in several parameters, all solutions on a connected component of the solution manifold are accessible if one solution on that connected component is known. But it might happen that we did not try that starting function (initial condition) which would have led to the discovery of another set of solutions, or that we miss a whole connected component.

The combination of analytical and numerical methods introduced by Schröder and Plum [Sc86, Plu91] promises to give more insight into the existence of solutions of underlying mathematical models in the future. At present, it is not applicable to systems of equations that are as involved as the ones investigated here.

Computations of Ken Meyer

Among the first numerical investigations of Taylor vortex flows were those by Ken Meyer [Mey66, Mey67]. He was very aware of the fact that the wavelength of the vortices is a behavior parameter in experiments and a control parameter in computations with periodic boundary conditions. He seemed to expect that the computations should be able to single out a *preferred wavelength* as the one that would occur in the experiment. This expectation seemed to be supported by his observations that his time-dependent computations converged to *stationary* Taylor vortex cells *of equal size* only for certain wavelengths. As an example we show in Fig. 3.2 his computations for $\mu = 0$, $\eta = 0.8333$, $\mathrm{Re} = 3.79\ \mathrm{Re}_{cr}$ and two different periods: in one case ($\lambda = 2.5$), the different vortices clearly have different axial lengths, while in the other case ($\lambda = 2.1$), the axial lengths do not differ so clearly. He reported that all flows had a tendency to move in the axial direction with time instead of becoming stationary.

By the standards of 1966, Ken Meyer did very fine, pioneering computations. With today's knowledge, we have to criticize his computations:

1) He did impose periodic boundary conditions on the velocity components, but he did *not* require a normalizing condition such as

$$w(r_o, z_o) = 0 \tag{3.1}$$

in at least one grid point (r_o, z_o) on the axial boundaries of his computational domains. Since every solution of the equations with periodic boundary conditions gives rise to a continuum of solutions shifted in the axial

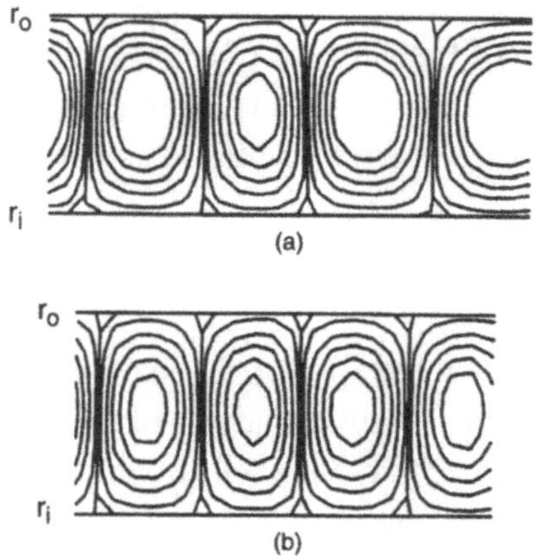

Figure 3.2: Computations with $\mu = 0$, $\eta = 0.8333$, $\mathrm{Re} = 3.79\ \mathrm{Re}_{cr}$ and the wavelengths (a) $\lambda = 2.5$ and (b) $\lambda = 2.1$ by Ken Meyer [Mey67, Fig.1].

direction, computations of such solutions will reliably converge only if one of them is singled out by *some* sort of normalization. In the numerical approach of Tayperio, this is obvious because the linear system solved by Gauss elimination has to be nonsingular for almost all parameter values. We thus imposed the corresponding condition on w in (2.3) already when we set up the system to be solved. For time dependent computations, this is not so obvious. It was discussed in detail by Booz [Boo, p. 58ff], who did computations both with and without condition (3.1). Booz found that he needed condition (3.1) to keep the solutions from moving in the axial direction with a speed depending on the grid. He also investigated the shape of interfaces between vortices in a few cases.[3]

2) Most of Meyer's computations were done with very crude discretizations (square meshes with 9 to 14 grid points across the gap width). The coarser a grid, the more important it is that the grid does have the same symmetry properties as the solution that is to be approximated.

We shall show in this chapter that Taylor vortex flows with non-equally sized vortices do exist and sometimes are stable. Thus further research showed that Meyer's concept of 'preferred wavelengths' was not adequate

[3]Later investigations with larger periods showed that interfaces can have shapes very different from those essentially plane interfaces Booz found, see for instance Fig. 3.13 or Fig. 3.28.

the way he formulated it. It should be noted, however, that his impressive and clear formulation of the concept of a 'preferred wavelength' was helpful for finding the correct answers to the questions of existence and stability of stationary Taylor vortex flows of different wavelengths.

Stability of Taylor vortex flows

The marginal curve at which Taylor vortex flows bifurcate from Couette flow has been found by a stability analysis of Couette flow. Only a stability analysis for Taylor vortex flows can tell in what parameter regimes Taylor vortex flows are stable. A first stability analysis for wide-gap Taylor vortex flows was performed by Kogelman/DiPrima [KD] by a perturbation analysis up to 3rd order. This analysis was done under several assumptions; for instance they assumed that Taylor vortex flows exist everywhere above the neutral curve, and nowhere below it. The marginal curve obtained by their analysis stays close to the marginal curve of Couette flow. It is discussed in the section on stability and is included in the figures giving those final theoretical stability limits that were confirmed by experiments; see Figs. 3.37ff.

Waves on deep water

In their investigations of waves on deep water, Chen and Saffman [ChS] found bifurcations halving (or doubling) the spatial wavelengths of stationary waves. Thus Phillip Saffman raised the question if such bifurcations between solutions with different spatial wavelengths are a general feature of the Navier-Stokes equations and could also be observed in different situations. This question was positively answered in the course of the research to be described next.

3.3.2 Continuous change of period

To better understand why wavelengths from a wide λ interval led to stable flows [KD], but why not all wavelengths inside the neutral curve of Couette flow seem to do so, and to see how flows of different wavelengths differ from each other, we systematically varied the period in computations with all other parameters fixed. We found a change of the periodic structure of the solutions, fold points, and period-changing bifurcations.

In the course of these investigations, we considered it *advantageous* that our numerical approach allowed us to treat the wavelength as a control parameter rather than as a behavior parameter: this made it easier to analyse the period-changing bifurcations and thus to understand why non-equally sized vortices exist.

Our first computations for fixed Re and varying period λ were performed in 1978 and are reported in [MK78, Table 4.2 and Fig. 4.9]. The first set of these computations was performed for $\eta = 0.95$, $\mathrm{Re} \approx 1.11\ \mathrm{Re}_{cr}$ and $N = 4$, $M = 31$. At that time the code was only set up for pseudo-arclength continuation in the Reynolds number Re for given period λ. Thus using one computed solution with $(\lambda_1, \mathrm{Re}_1)$ as start function for a solution with $(\lambda_2, \mathrm{Re}_1)$ resulted in a solution with $(\lambda_2, \mathrm{Re}_2)$, $|\mathrm{Re}_1 - \mathrm{Re}_2|$ small. Starting with a flow at $\lambda = 2.007$ and $\mathrm{Re}^\delta = 205.4$, we ended at Couette flow with $\lambda = 1.317$ and $\mathrm{Re}^\delta = 203.2$. Six solutions were thus computed in the parameter range $1.317 \leq \lambda \leq 2.007$; $203.2 \leq \mathrm{Re}^\delta \leq 205.4$. While the maximum and minimum of the radial velocity component $u(r, z)$ did not change noticeably for fixed $\lambda = 2.007$ and $201 \leq \mathrm{Re}^\delta \leq 205.4$, both $\min u(r, z)$ and $\max u(r, z)$ changed to the Couette value 0 with λ. The minimum $\min u_{\min} = -2.5\ 10^{-2}$ was observed at $\lambda = 1.8007$, the maximum $\max u_{\max} = 3.3\ 10^{-2}$ at $\lambda = 2.007$. The occurrence of the largest return flow from the outer to the inner cylinder $(u(r, z) < 0)$ for $\lambda \approx 1.8$ is in agreement with the experiments of Donnelly and Schwarz for $\mathrm{Re}^\delta \approx 200$ as reported in [Mey67, p. 1878]. The return flow back to the inner cylinder was weaker than the flow from the inner to the outer cylinder for all periods checked. This same lack of symmetry could be observed in many other computations, for instance in those leading to Fig. 3.4 for $\lambda \geq 1.6$. Because the flow is divergence free, the spatial range of the return flow must be larger than the spatial range of the outward flow, as in Fig. 3.1b and in Fig. 2.8.

For similar computations with $\eta = 0.5$ the code was extended to allow for simple continuation in the period with fixed Reynolds number. Computations with fixed $\mathrm{Re} = 101.8 \approx 1.5\ \mathrm{Re}_{cr}$ were done for 8 values in the interval $1.288 \leq \lambda \leq 1.988$. It was found that the torque had a maximum at about $\lambda = 1.688$: 5% higher than for Couette flow ($\lambda = 1.288$) and 1.7% higher than for $\lambda = 1.988$. Such maxima of the torque were also calculated by DiPrima/Eagles [DPE] up to $\mathrm{Re} = 79.4 \approx 1.17\ \mathrm{Re}_{cr}$. A more complete picture was later given by Riecke, Paap: they plotted the possible maximum and minimum values of the torque together with the values for λ_{cr}

and the measurements of [DS] in the Re-interval $\mathrm{Re}_{cr} \leq \mathrm{Re} \leq 2\,\mathrm{Re}_{cr}$ [RP86, Fig. 1]. They did not report the wavelengths at which the maximum and minimum occur: since the concept of a 'preferred wavelength' proved to be wrong, those wavelengths maximizing or minimizing the torque cannot be the preferred ones.

Using simple continuation in the period as just described, we [FM] then computed flows for $\eta = 0.5$ and varying periods larger than two. Both when doing continuation in the Reynolds number (fixed period) or simple continuation in the period (fixed Reynolds number), we encountered parameter intervals in which the number of vortices changed smoothly (from 2 to 4, from 2 to 6), and also parameter values for which Newton's method did not converge, and where the Jacobian became singular. The points of non-convergence were detected as fold points of solution branches. These phenomena were observed close to intersection points of the neutral curve of Couette flow with its periodic repetitions, as shown in the sketches in Figs. 2.7 and 3.3. A large part of this and the next chapter will be devoted to explaining these findings and to deducing consequences.

Though the computations of [FM] were handicapped by the fact that simple continuation cannot treat fold points in a satisfactory way, these computations reliably showed that flows with vortices of different sizes do exist and that they are not numerical artifacts due to computations with 'wrong' (not preferred) wavelengths or wrongly spaced grids. Many of the flows we found were intermediate states between solutions with equally spaced vortices (transitions from 2 to 4 and from 2 to 6 vortices). As an example we repeat here [FM, Fig. 8] as Fig. 3.3. It shows two flows close to the $(2,6)$-intersection point of the neutral curves. Note that there are two vortices contained *inside* of one vortex. Also note the similarity between this flow and flows shown in Figs. 3.32 and 4.13.

These computations also revealed that bifurcation from Couette flow to Taylor vortex flows is sometimes subcritical. It had generally been assumed that the convex (parabolic shaped) neutral curves indicating bifurcations from Couette flow separate the (Re, λ)-plane into regions where $2n$-vortex solutions exist (above) and where they don't exist (below). This assumption is correct for the highly symmetric Rayleigh-Bénard problem (see next chapter). For the wide gap Taylor problem, it has proven to be incorrect. We found a λ-interval to the right of the intersection of the neutral curves for 2 and 4 vortices in the (Re, λ)-plane where the bifurcating Taylor vortex flows exist for Reynolds numbers *below* the critical Re-value for that λ.

This is in agreement with the analysis of Andreichikov [An77] and does not contradict the analysis of Ovchinnikova, Yudovich 1974 [OvY] and Stuart, DiPrima 1980 [SD80].

Andreichikov [An75, An77] investigated the 2-4 bifurcations in a neighborhood of the intersection of the two neutral curves for 2 and 4 vortices. This analysis is given in the case of periodic boundary conditions and no flow across the boundary, for general η. The general formulas obtained by calculations were then numerically evaluated in the case of $\eta = 0.5$ which we consider here. Thus Andreichikov's results are directly applicable to the problem considered here and inspired the investigations to be discussed next. His results for $\eta = 0.5$ are summarized in [An77, Fig. 4]. For easier reading of this figure we modified it (wavelength instead of wave number) and repeated it as [FM, Fig.7]. In this text it is Fig. 3.14. From Andreichikov's analysis it is also clear where subcritical bifurcations exist for $\eta = 0.5$, see Fig. 3.14.

Assuming periodic boundary conditions in the axial direction, Ovchinnikova and Yudovich proved that the bifurcation is supercritical for all wavelengths in the limiting case $\eta \to 1$ and $\mu \to 1$. In our case $\mu = 0$ and $\eta = 0.5$, and thus their analysis is not applicable. Stuart and DiPrima assumed different boundary conditions in the axial direction. They formulated conditions under which the bifurcation is respectively supercritical or transcritical, but only for narrow gaps ($\eta \approx 1$). In the limiting case $\eta \to 1$ and $\mu \to 1$, they again proved that the bifurcation is supercritical for all wavelengths.

As an example for the computations of [FM] we show Fig. 3.3. Though these computations were done with very coarse discretizations, they do give qualitatively correct results: see the section on numerical accuracy in Chapter 2 for more details.

3.3.3 Flows near Re $= \sqrt{1.5}$ Re$_{cr}$

We now study in more detail the structure of the families of solutions (or flows) with respect to continuous variation of λ and Re. As gap width for these investigations we chose $\eta = 0.727$, as Reynolds number we first chose Re $= \sqrt{1.5}$ Re$_{cr}$.

While Taylor vortex flows are known to bifurcate 'very soon' to wavy vortices in the narrow gap case (for Re/Re$_{cr} < 1.2$ if $\eta = 0.95$), they stay axisymmetric for a large range of Reynolds numbers in the wide gap case. We chose to do further research on stationary vortices of changing

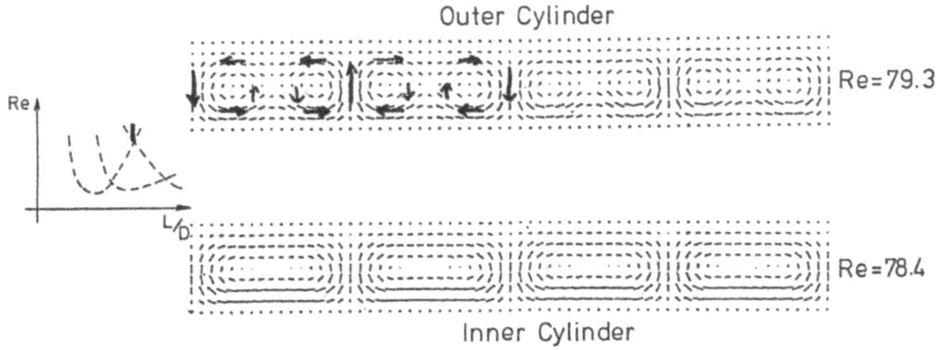

Figure 3.3: The (u, w) vector field in a 2λ long axial section of the gap is shown for $L/D = \lambda = 3.7$, $\eta = 0.5$, $\mathrm{Re} = 78.4$ and $\mathrm{Re} = 79.3$.
The arrows mark the loci of the maxima of u and w, respectively. These computations used $N = 4$ modes and $M = 7$ grid points [FM, Fig. 8].

wavelengths for $\eta = 0.727$, which is a 'narrow wide gap' according to Jones' stability analysis (see Fig. 3.35), and for which stationary vortices of various wavelengths were observed in the experiments of Burkhalter/Koschmieder up to $\mathrm{Re}/\mathrm{Re}_{cr} > 9$ (remember Fig. 1.8 and earlier comments about it).

The value $\mathrm{Re} = \sqrt{1.5}\,\mathrm{Re}_{cr}$ was chosen for these investigations both for mathematical and physical reasons:

1) The investigations of Andreichikov[An75, An77] and Frank, Meyer-Spasche [FM] discussed earlier showed that intersection points of neutral curves influence the existence of solutions. To optimally continue those investigations, it seemed desirable to be 'well above' the first of these intersection points, but not above the second of these points, so as not to have a mixup of too many effects. Thus the Re-value of the second intersection point seemed to be optimal for $\eta = 0.5$ [FM, Figs. 3 - 8]. We expected similar things to happen in the case of $\eta = 0.727$, in the neighborhood of the corresponding intersection points. This proved to be correct.

2) In the experiments of Burkhalter, Koschmieder [BuKo], the range of observed wavelengths is quite wide in a neighborhood of this Re-value: They generated Taylor vortex flows with wavelength $\lambda = 1.5$ by sudden starts, and Taylor vortex flows with $\lambda = 2.106$ and $\lambda = 2.108$ by filling experiments, both for Re-numbers larger and smaller, but always close to $\mathrm{Re} = \sqrt{1.5}\,\mathrm{Re}_{cr}$ (see [BuKo, Figs. 7, 9 and 10], and Fig. 1.8).

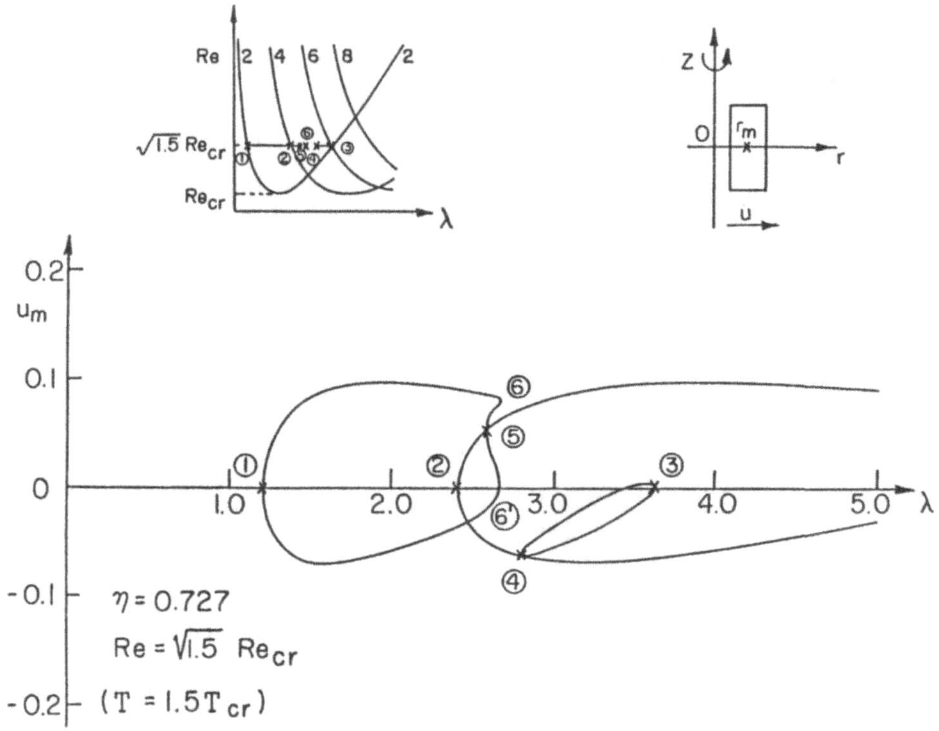

Figure 3.4: Variation of the radial velocity u_m at the midpoint with the period λ. Points (1), (2), (3): Bifurcation of 2-vortex flow and two periodic repetitions from Couette flow ($u_m \equiv 0$). Point (3) is a bifurcation point of codimension 2, a 2-vortex flow bifurcates there as well, see Fig. 3.7. Points (4), (5): period-halving bifurcation points. Fold points in branches of 2-vortex flows occur at (6) and (6'). Computations with M = 31 interior radial grid points and N = 6 Fourier modes [MK85, Fig.1].

λ-continuation paths

The solutions of eqs. (2.1)–(2.3) are elements of a function space of infinite dimension. Our numerical approximations are elements of some $I\!R^{(2N+1)(2M+1)}$ where N and M are the number of grid points and of Fourier modes, resp.. As soon as we found bifurcations between solution branches, we found it difficult to display the rich variety of computed solutions and their relations to each other in simple plane graphs. The torque graphs usually given in studies of Taylor vortex flows do not show interactions of solution branches that turned out to be important. An alternative choice is plotting a velocity component, for instance the radial velocity at the center of the computational domain, $u_m := u(r_m, 0)$ vs. the parameter varied. Positive u_m corresponds to outward directed flow. We also present a graph of poloidal kinetic energy versus λ. This is similar to the torque graphs.

The main results of this section are summarized in Fig. 3.4 where u_m vs λ is plotted for fixed $\eta = 0.727$ and $\mathrm{Re} = \sqrt{1.5}\ \mathrm{Re}_{cr}$. We now describe how the numerical continuation and bifurcation techniques described in Chapter 2 were used to obtain these results. Starting with Couette flow at $\lambda = 2$ and $\mathrm{Re}_{cr} \approx 219.10$ (the numerically found approximation to the bifurcation point), branch switching was used to compute the Taylor vortex flows bifurcating from Couette flow for increasing Re-values. Using pseudo-arclength continuation in Re, this branch was then continued up to $\mathrm{Re} = \sqrt{1.5}\ \mathrm{Re}_{cr}$. Actually only one half-branch was computed since the other half-branch can be obtained from it via a shift in the z coordinate by π/k: it is a wavelength-shifted twin as introduced in Chapter 2. (The bifurcation from Couette flow is of the 'supercritical' or 'pitchfork' type, see Fig. 2.4).

Starting from one of the two bifurcated Taylor vortex flows at $\lambda = 2$, $\mathrm{Re} = \sqrt{1.5}\ \mathrm{Re}_{cr}$, we then did simple continuation in λ, using the values $\lambda = 2(-.05)1.2$ for decreasing λ and $\lambda = 2(.05)2.65$ for increasing λ.

The decreasing λ continuation smoothly switches back to Couette flow at $\lambda = 1.2$ since there is apparently no other neighboring solution (at the point labeled (1) in Fig. 3.4). The amplitude of the vortices decreases smoothly to zero. This can be seen from Fig. 3.5a showing the poloidal kinetic energy E_{pol} versus λ.

The increasing λ continuation, however, failed to converge at $\lambda = 2.70$. That such points exist was already known from the investigations of [FM] discussed in the previous section and led to the implementation of pseudo-arclength continuation with respect to $k = \frac{2\pi}{\lambda\delta}$. This procedure was used now, starting at $\lambda = 2.65$. A fold point was found at $\lambda \approx 2.6754$ (the point labeled (6) in Fig. 3.4). Before this fold was reached the flow changed its structure: two additional vortices were formed where the flow towards the inner cylinder had slowed down. As this path of solutions is traversed, the new small, weak vortices grow in magnitude and strength and the flow changes continuously to a 4-vortex flow with equally sized vortices. How this change develops is illustrated by Figs. 2.7 and 3.27. Though these plots were done for different parameter values, the pictures for the case considered here were similar. Even the streamlines displayed in Jones [Jo82, Fig. 3] are very similar, though they deal with the case of counter-rotating cylinders (Re varying, $\lambda = 2,\ \eta = 0.5,\ \mu = -0.2$).

Upon passing around the fold the continuation procedure decreases λ and near $\lambda = 2.59$, labeled (5) in Fig. 3.4, the solution joins smoothly to the other half-branch of the Taylor vortex flows which bifurcated from Couette flow at (1). This follows from a smooth sign change in all of the odd Fourier coefficients at all of the mesh points. This sign change is equivalent to a shift along the z-axis by the amount π/k, i.e. to its $\lambda/2$-shifted twin (see the section on 'periodicity' in Chapter 2). So in effect we have determined a closed loop or family of solutions, parameterized by λ, with fixed η, μ and Re. This loop contains the points (1) $\to$ (6) $\to$ (5) $\to$ (6') $\to$ (1) in Fig. 3.4. It is not symmetric with respect to the λ-axis. This was surprising at first. It will be explained later, when the relation to the Rayleigh-Bénard problem will be discussed.

Point (5) does not only join the two half-branches. It also lies on the $\times 2$-branch of 4-vortex solutions that bifurcates from Couette flow at point (2): bisection iterations in λ for the sign change of one of the Fourier coefficients confirmed that the sign change happens simultaneously in all the odd Fourier coefficients at all the mesh points. This proves that each odd Fourier coefficient vanishes at point (5) and that point (5) also belongs to the $\times 2$-branch. Thus we have found a secondary bifurcation between a 2-vortex branch and a 4-vortex branch. We stress, however, that neither branch *remains* a pure 2- or 4-vortex branch; they are named for the flow that develops when they bifurcate from Couette flow.

At the point labeled (3) a multiple bifurcation from Couette flow occurs; both 2-vortex and 6-vortex flows bifurcate there. The branch for 2-vortex flows exists for decreasing λ-values, but changes continuously into a 4-vortex flow. This time, however, the additional vortices form where the flow is outward. Continued decrease of λ yields strengthened vortices, and eventually this branch joins its $\lambda/2$-shifted twin. In the same way as before, it was confirmed by bisection that (4) is a secondary bifurcation point. We thus found another closed loop of solutions joining the bifurcation points (3) and (4). Streamlines of the changes that occur along such a solution branch can be seen in [FM, Fig. 5] and in Fig. 3.28 a,b here. In Fig. 3.6 we illustrate schematically both changes (1)–(6)–(5) and (3)–(4).

Let us stress again that we confirmed by detailed tests:

The change of structure of the solutions caused by the generation of the additional vortices happens continuously. It does not start at another bi-

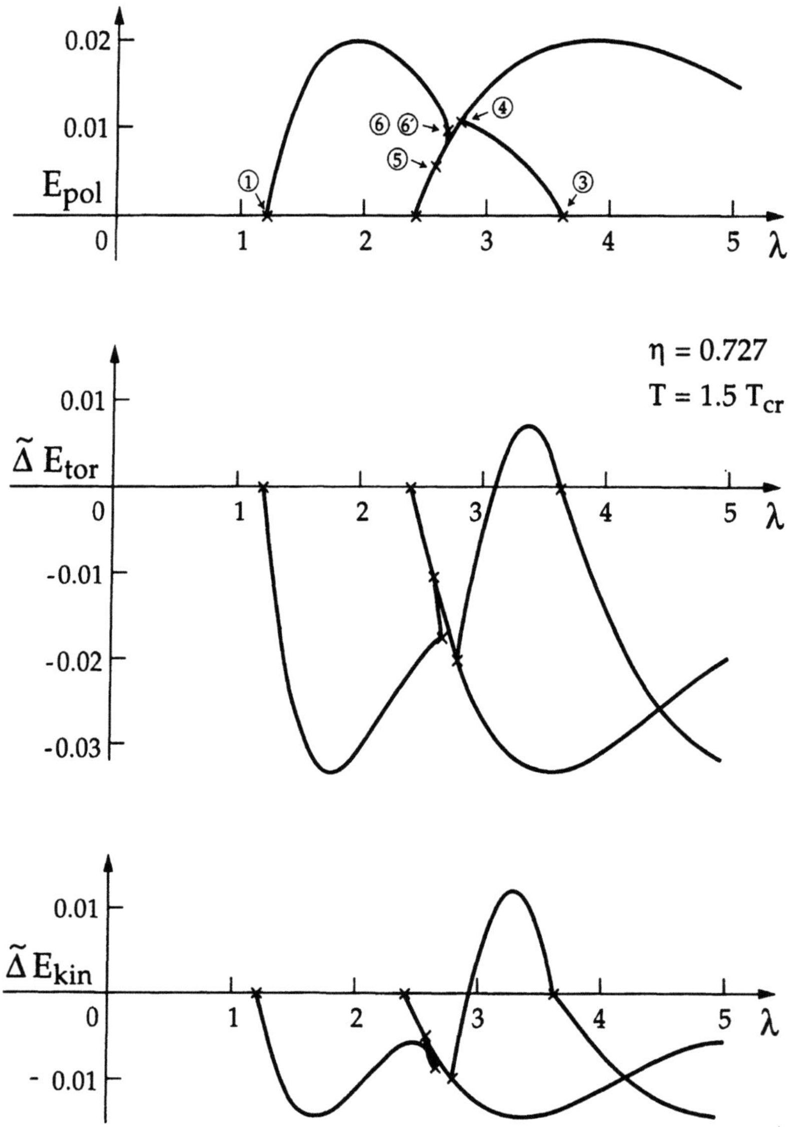

Figure 3.5: Kinetic energies versus period λ.
a) Poloidal kinetic energy E_{pol} versus period λ. The same branches and critical points as in Fig. 3.4 are shown, but $\lambda/2$-shifted twins coincide here. A torque versus λ graph would look very similar [MK85, Fig.3].
b) $\tilde{\Delta}E_{tor} := E_{tor}(\text{Taylor vortices})/E_{tor}(\text{Couette flow})-1$ vs. λ. At negative values, E_{tor} of the Taylor vortices is smaller than that of Couette flow, at positive values it is larger.
c) $\tilde{\Delta}E_{kin} := E_{kin}(\text{Taylor vortices})/E_{kin}(\text{Couette flow})-1$ vs. λ.

furcation point between (1) and (6) (or between (3) and (4)) nor at the fold points (6), (6').

At the bifurcation points (5) and (4) all odd Fourier coefficients change their sign *simultaneously*. Our numerical approximation does not perturb these bifurcations. This is due to the fact that we use periodic boundary conditions together with a Fourier decomposition. In other numerical approximations these bifurcations could get perturbed, see for instance [ScK, Figs. 6,7].

In Fig. 3.5 the branches of solutions described above are plotted in terms of their kinetic energies. Figure 3.5a shows the poloidal kinetic energy (2.104), i.e. the kinetic energy of the (u, w) vector field. The critical points (1), ..., (6') are indicated on the graph, for the reader's convenience. The $\lambda/2$-shifted twins have the same energy values and thus coincide here: important details are concealed. A torque versus λ graph would look very similar. More information on changes of solutions with λ is contained in Fig. 3.4 than in Fig. 3.5a.

The total kinetic energy E_{kin} versus λ and the toroidal kinetic energy E_{tor} were also computed and plotted. We found that the increase (decrease) of the poloidal energy E_{pol} often but not always coincides with a decrease (increase) both of E_{tor} and E_{kin}. The values of E_{tor} and E_{kin} are below their values for Couette flow on the whole branch (1)–(6)–(5). The maximum of E_{pol} and the minima of E_{tor} and E_{kin} all occur at $\lambda \approx 1.95$. At its maximum, E_{pol} contains about 0.96% of the kinetic energy of Couette flow, while E_{tor} contains 92.5% and E_{kin} contains $\approx 93.5\%$ there [MK84, p.28]. When starting at (3) with decreasing λ, E_{tor} and E_{kin} first grow above their values for Couette flow, but decrease when the little vortices are formed. Figure 3.5b shows $\tilde{\Delta} E_{tor} := E_{tor}(\text{Taylor vortices})/E_{tor}(\text{Couette flow})-1$ versus λ and Fig. 3.5c shows $\tilde{\Delta} E_{kin} := E_{kin}(\text{Taylor vortices})/E_{kin}(\text{Couette flow})-1$ versus λ, i.e. the relative change of kinetic energies due to the forming of vortices. – Both E_{pol} and E_{kin} are shown in Fig. 3.26. On the branches investigated there, the total kinetic energy of the vortices never exceeds that of Couette flow.

Figure 3.7 is an extension of Fig. 3.4, with more periodic repetitions. The proper range of possible periods is determined by the stability of the flows. This will be discussed in a later section, near Figs. 3.36 and 3.37.

As we showed earlier, there are some values of η and λ for which Taylor vortex flows exist *below* the neutral curve of Couette flow. Our current computations indicate that the converse may also be true. This is suggested

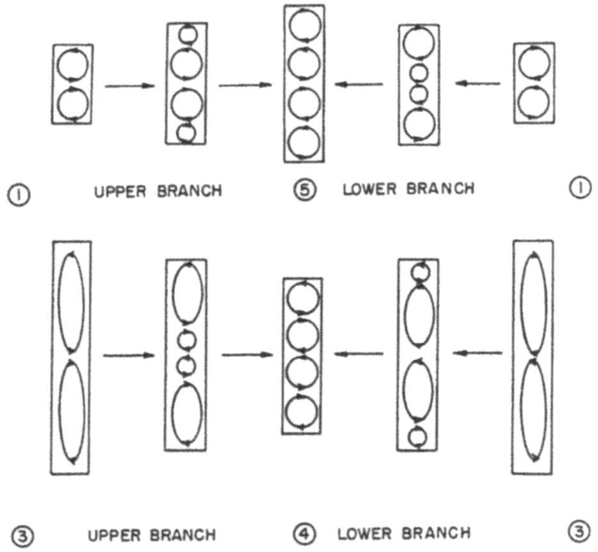

Figure 3.6: Schematic diagram of how the cellular structure of Taylor vortex flows changes with wavelength between 2-vortex and 4-vortex flows. The sketches indicate the transitions with λ along the upper and lower paths (1)–(5) and (3)–(4) of Fig. 3.4. Streamline patterns of computed transitions are shown in Figs. 3.27 and 3.28 a,b, respectively [MK85, Fig.2].

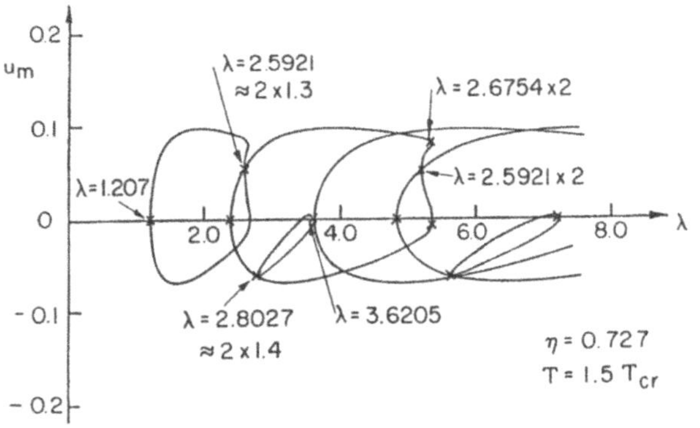

Figure 3.7: As in Fig. 3.4, but higher period branches are also shown. The multiple bifurcation at $\lambda^* := 3.6205$ contains Couette flow, 2-vortex flow for $\lambda \leq \lambda^*$ and 6-vortex flow for $\lambda \geq \lambda^*$: it is the $(2,6)$ double point. The experiments of Burkhalter/Koschmieder include $1.5 \leq \lambda \leq 2.108$ for the same η and an interval of T containing $T = 1.5T_{cr}$ [MK85, Fig.4].

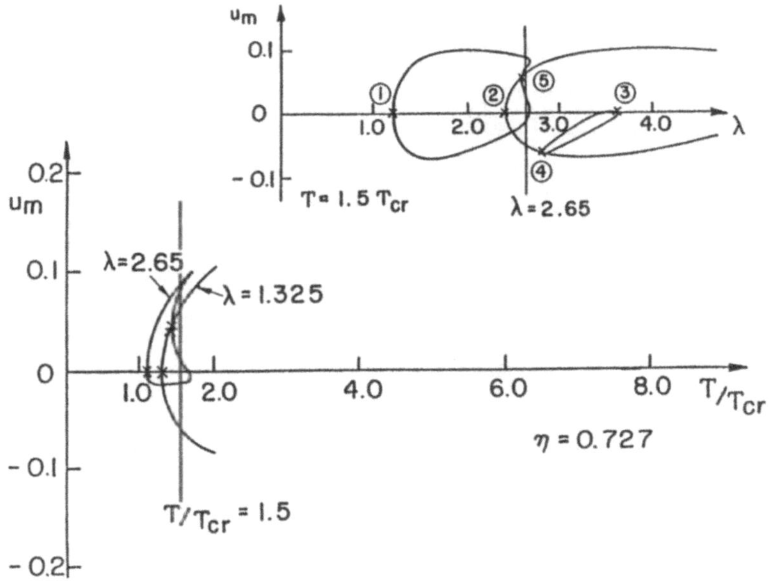

Figure 3.8: Variation of the radial velocity u_m at the midpoint with the Taylor number. The solution branch for $\lambda = 1.325$ has a spatial period doubling bifurcation to the branch for $\lambda = 2.65$ [MK85, Fig.5].

by the 'gap' in the λ-traverse shown in Fig. 3.4 (upper inset). For the λ interval between (6) and (4) no 2-vortex branch solutions were found. This will be discussed further when dealing with the 'interactions of secondary branches'.

Re-continuation paths

For the period $\lambda = 2.65$ we have done continuation with respect to Re, or rather, $T \sim \mathrm{Re}^2$ (see Fig. 3.8). This λ value is situated between the fold at (6), (6') and the secondary bifurcation at (5). As shown in Fig. 3.8 we found two folds with respect to T on the $\lambda = 2.65$ branch, and a spatial period-halving bifurcation to a $\lambda = 1.325$ branch. Very similar were the results for $\lambda = 3$ / $\lambda = 1.5$.

For other λ values, however, for instance for $\lambda = 2/\ \lambda = 1$, we found the period doubling bifurcation on (the periodic repetition of) the $\lambda = 1$ branch, but not the fold point; see Fig. 3.22. This agrees with the computations of Jones [Jo82] for $\eta = 0.8756$. It will be explained later (near Fig. 3.23).

Experimental results

The reported computations were done on 'infinite' cylinders, while all experiments have to be done in finite cylinders with real end conditions. As already mentioned, Burkhalter and Koschmieder [Bu72, BuKo] explicitly studied the effect of end conditions on Taylor vortex flows. They conducted experiments with rotating solid end boundaries, non-rotating solid end boundaries, and free surface on top. Also, they varied the lengths of the cylinders. In all of their experiments they measured both the lengths λ_E of the Ekman vortices at the ends and the averaged length λ_f of the inner cells. They concluded:

If Re is increased quasisteadily, changes in the wavelength λ_f of the inner vortices are caused by end effects. An extrapolation to infinitely long cylinders gives constant wavelength $\lambda_f \approx \lambda_{cr}$.

In sudden start experiments, shorter wavelengths $\lambda_f < \lambda_{cr}$ are preferred and λ_f clearly depends on Re.

When they changed the length of the fluid column by filling the fluid while the inner cylinder was rotating, they observed the creation of new pairs of cells, but $\lambda_f \geq \lambda_{cr}$.

Our computations suggest the following interpretation of these results:

If Re is increased quasisteadily, Couette flow first becomes unstable at $\mathrm{Re} \approx \mathrm{Re}_{cr}$, and Taylor vortices form with $\lambda_f \approx \lambda_{cr}$ (λ_f equal λ_{cr} up to the quantization error (discretization error) introduced by the fact that the vortices have to fit into a given apparatus). As long as (λ_f, Re) or (λ_E, Re) do not enter the parameter regime where transitions occur, the wavelength cannot change.

In the sudden start experiments, the flow is dominated by the growth of the boundary layer on the impulsively rotated cylinder. The temporal development of such a flow is displayed in four photographs produced by Taneda [Dy, Fig. 121]. Small vortices are generated along the full length of the inner cylinder simultaneously, and then grow into the volume. Since they start as very small vortices in the boundary layer, it is not surprising that the final stationary vortices have wavelengths λ_f smaller than λ_{cr}. Our computations show a unique relation between the kinetic energy of stationary vortices and their wavelength for $\lambda_f < \lambda_{cr}$. This suggests that the time-dependent vortices have to grow in size because their kinetic energy grows.

In the filling experiments, the wavelength grows until it comes to the regime where additional vortices form, it decays, grows again, ... Thus it is plausible that in this case $\lambda_f \geq \lambda_{cr}$.

In our computations, we found transitions from 2 to 4 and 2 to 6 [FM] vortices. In a long cylinder, that would correspond to transitions from 10 to 20 vortices or so. As far as we know, such things have never been observed in experiments. Usually, one new pair of vortices forms or disappears [Bu72, Co65, AC]. The transitions 2 to 4 just discussed explain the fundamental mechanism; whether they also describe a transition that can actually occur in real flows seems reasonable at present: see the discussion of the short-wavelength instability later in this chapter, near Fig. 3.40. Riecke, Paap [RP90] consider it possible that the flow finally resulting from the short-wavelength instability is a stationary flow with twice the wavelength of the initial flow because every second vortex pair disappeared.

In his very short apparatus, Benjamin [Ben78] did observe transitions from 2 to 4 vortices. Also, he observed that the branch of two vortices undergoes a fold. Surprisingly enough, it is possible that the main effect of his cylinders being so short is a stabilization of flows that are unstable in longer cylinders.

3.3.4 Flows for Re $= 800 \approx 3.65$ Re$_{cr}$

In the previous section we discussed the structure of families of flows for Reynolds numbers close to Re $= \sqrt{1.5}$ Re$_{cr}$. In a paper following the original study of Meyer-Spasche and Keller [MK85], Dinar and Keller [DK] investigated the dependence of fold point (6) on the period λ and on Re. They showed that the fold point encountered for fixed λ and varying Re is a point of type (6) when approached by varying λ for fixed Re. Thus the two different fold points shown in the two parts of Fig. 3.8 both belong to the same fold, which we shall call 'the basic (2,4)-fold'. We shall discuss it in detail in the next section. In this section we discuss flows which Meyer-Spasche and Wagner [MW1] found when investigating the basic (2,4)-fold. These flows were computed for varying period λ and fixed values Re $= 800 \approx 3.65$ Re$_{cr}$, $\eta = 0.727$, $\mu = 0$. Some of these flows have quite unusual structures: There are *double vortex flows* with two adjacent vortices in the radial direction and *free stagnation points of the poloidal flow.* And then there is 'flow # 13' mentioned earlier, Fig. 3.13.

Figure 3.9a shows the branch which we obtained by continuation in λ when starting with the usual 2-vortex solution for Re $= 800$, $\lambda = 2$.

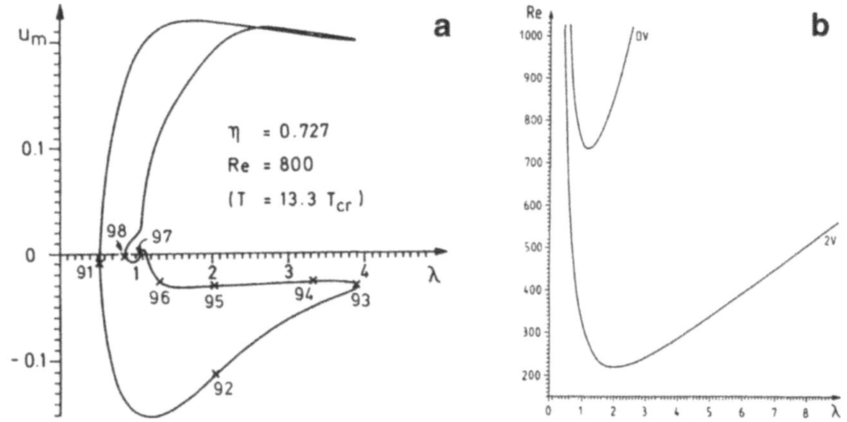

Figure 3.9: a) Basic 2-vortex branch for Re $= 800$, $\eta = 0.727$, $\mu = 0$. Bifurcation of 2-vortex flow from Couette flow at $\lambda \approx 0.52$, fold point at $\lambda \approx 3.9$, bifurcation of double-vortex flow to Couette flow at $\lambda \approx 0.85$.
b) Curves of bifurcation from Couette flow. 2V: 2-vortex flow. DV: double-vortex flow [MW1, Fig.1].

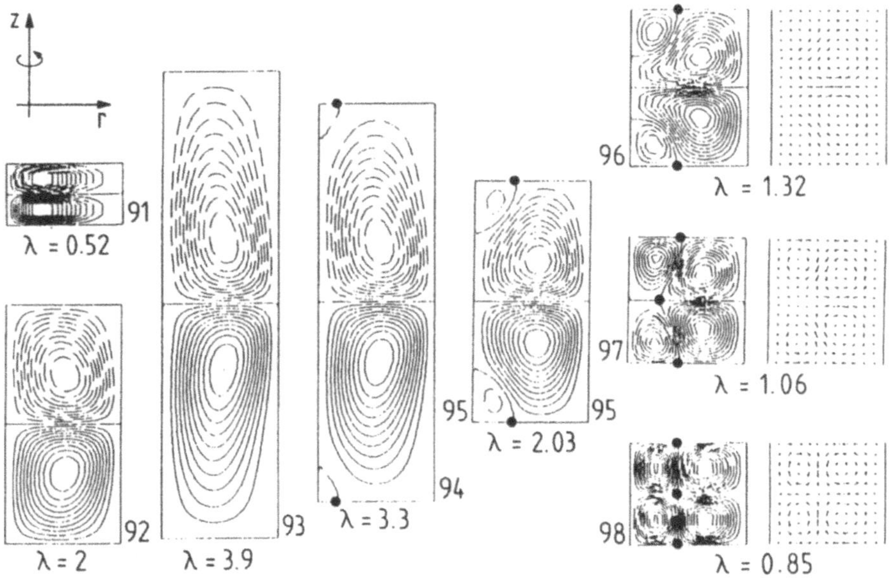

Figure 3.10: Streamline patterns of some of the flows on the basic branch. The flows shown are the $\lambda/2$-shifted twins of the flows marked in Fig. 3.9a. •: free stagnation point. [MW1, Fig.2].

As the computations showed, it could also be obtained through bifurcation from Couette flow at $\lambda \approx 0.52$. This is the only complete branch which we display for Re $= 3.65$ Re_{cr}. The computation of all other branches was interrupted before it came to a 'natural' stop, partially because of numerical difficulties (to be discussed in connection with Figs. 3.21 and 2.10), partially because we felt that things became too complex to give any further clues. The computations for the smaller value Re $= 2$ $\text{Re}_{cr} \approx 438$ were performed in the whole λ-interval given by the neutral curve for Couette flow. That promised to give more insight, and actually did.

Figure 3.10 shows the streamline patterns of some flows on the branch shown in Fig. 3.9a. Flows with 2 vortices in the radial direction are well-known for counterrotating cylinders. The existence of such flows in the case when the outer cylinder is at rest is not surprising: they bifurcate from Couette flow at a singularity of the Jacobian of system (2.1)–(2.3). This singularity corresponds to a higher (real) eigenvalue of system (1.35). We display the locus of this singularity in the (Re, λ)-plane in Fig. 3.9b, together with the common neutral curve (first eigenvalue). The flows bifurcating from Couette flow with two vortices in the radial direction (flow # 98) are called **double-vortex flows**. The flows bifurcating from Couette flow with three vortices in the radial direction are called **triple-vortex flows**. Their curve of bifurcation from Couette flow is shown in Figure (4.14), together with one periodic repetition.

A flow around an obstacle often has a stagnation point $\mathbf{v} = \mathbf{0}$ somewhere at the surface of the body. Stagnation points not situated at a rigid boundary but in the interior of a flow are called **free stagnation points**. They are not that common. There are cases in which their existence was an open mathematical question for many years. The stagnation points encountered here (see for instance Fig. 3.10) are not proper free stagnation points: the poloidal components of the velocity vanish at these points, but not the toroidal component. They are **'free stagnation points of the poloidal flow'**. Many of our figures display poloidal flows only. In the captures of those figures we call them 'free stagnation points' and omit 'of the poloidal flow'.

In Fig. 3.11 we display a second branch of solutions for Re $\equiv 800$ in three different projections, and also a sketch of it. In the parameter regime investigated, this second branch does not bifurcate from Couette flow. We happened to find it when we repeatedly performed the computations leading to Fig. 3.9a in a neighborhood of the fold point, with different step sizes in the pseudo-arclength, (2.42). As was confirmed by very detailed and very

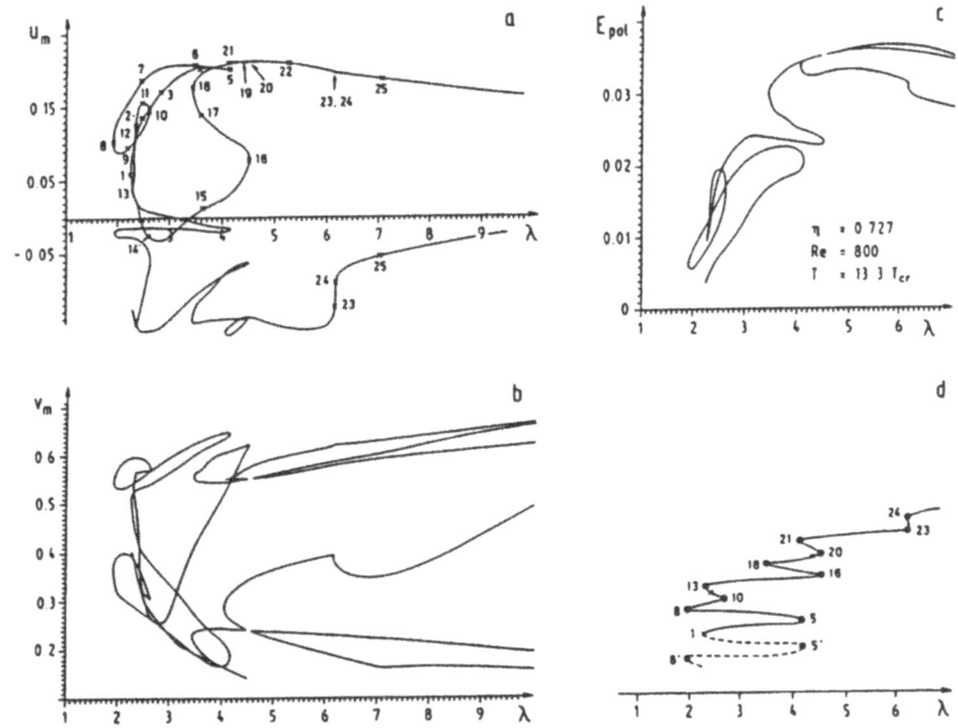

Figure 3.11: Second Re = 800-branch in different projections a) radial velocity u_m versus λ, b) toroidal or azimuthal velocity v_m versus λ, c) poloidal kinetic energy E_{pol} versus λ, d) sketch. This branch has 21 fold points and 3 bifurcation points. A third solution branch is included in parts b) and c), but not in parts a) and d) of this figure [MW1, Fig.3].

careful tests, there are two adjacent branches of solutions (to be discussed further near Fig. 3.20). How these branches are connected was found to depend on the discretization used (to be discussed near Fig. 2.10). In Fig. 3.12 we display how the structure of the flow changes along this branch when the period λ is varied continuously. More details are given in the caption. The strangest flow is "# 13", which seems to contradict the fact that the vortex movement is caused by centrifugal forces. We thus give more details about this flow in Fig. 3.13.

The picture given by Figs. 3.9 and 3.11 is still not complete: no periodic repetitions of branches are plotted, and there is at least one more original branch, separated from the main branch of Fig. 3.11 by a per-

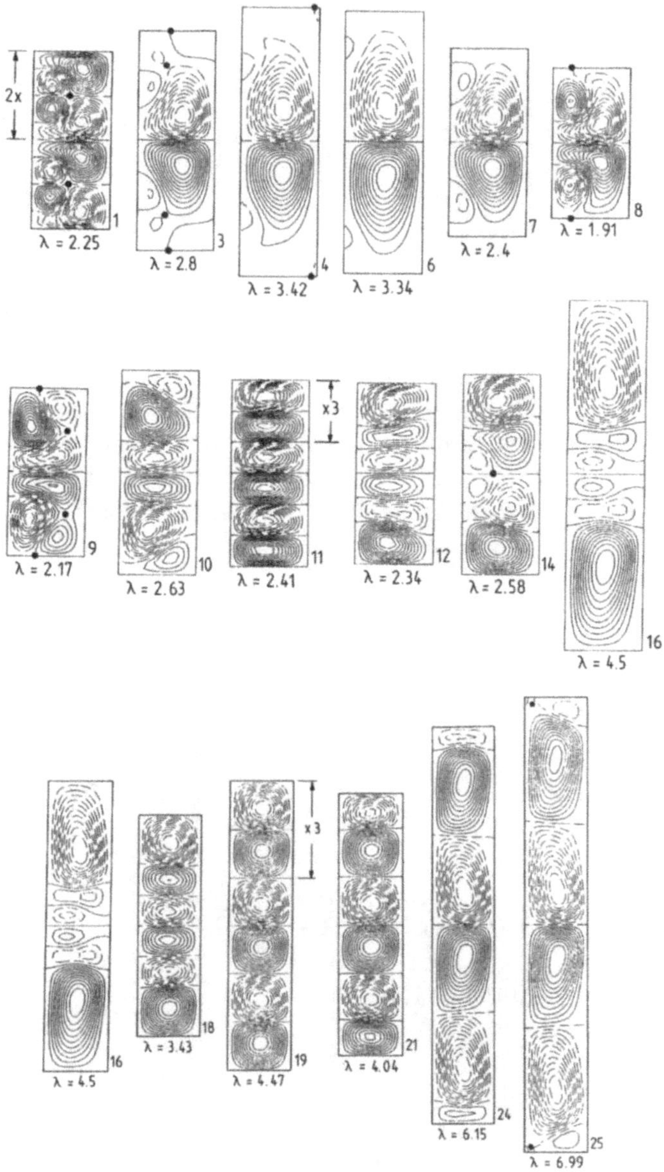

Figure 3.12: Change of streamline patterns with λ on the second Re = 800 branch. Numbers to the right coincide with numbers in Fig. 3.11. Factors are given for the change of period at the 3 bifurcation points. •: free stagnation points [MW1, Fig.4].

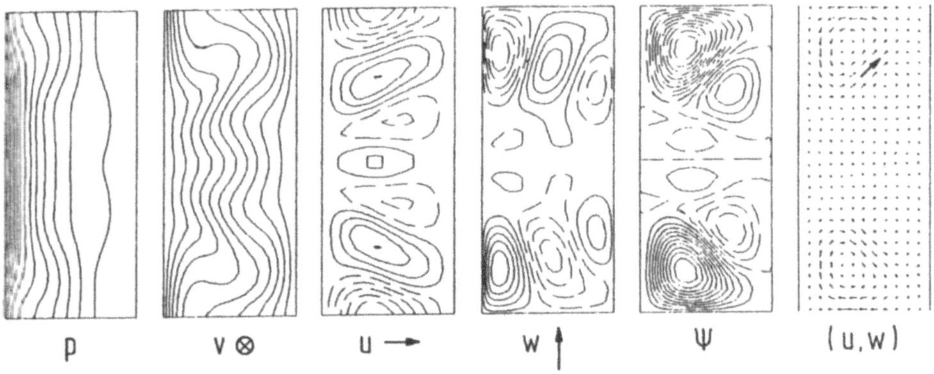

Figure 3.13: Level lines of pressure p, velocity components v, u, w and stream function ψ for Re = 800, $\eta = .727$, $\lambda = 2.27$ (flow '# 13', fold point in Fig. 3.11). For this flow, the main outward movement is found in diagonal directions in (r, z) [MW1, Fig.5].

turbed bifurcation and included in parts b) and c) of the figure. For the values $\lambda = 2.5$, Re = 800, $\eta = 0.727$ Figs. 3.9 and 3.11 tell us that there are at least 21 different solutions: 10 flows with minimal period 2.5 (Fig. 3.11d) *and* 4 flows with minimal period 2.5, 4 flows with minimal period $2.5/2 = 1.25$, and 2 flows with minimal period $2.5/4 = 0.625$ (Fig. 3.9), and Couette flow, of course.

3.4 Interactions of secondary branches

3.4.1 A neighborhood of $(\text{Re}_{24}, \lambda_{24})$ and the basic $(2,4)$ fold

The curve of neutral stability of Couette flow in the (Re, λ)-plane with minimum $(\text{Re}_{cr}, \lambda_{cr})$, $\lambda_{cr} \approx 2$, intersects with its periodic repetition (defined in Chapter 2) with minimum $(\text{Re}_{cr}, 2\lambda_{cr} \approx 4)$ at a point $(\text{Re}_{24}, \lambda_{24})$[4]. On both neutral curves, the Jacobian of the system (2.1)–(2.3) is singular, and at $(\text{Re}_{24}, \lambda_{24})$ it is singular of order 2. On both neutral curves, Taylor vortex flows bifurcate from Couette flow, and their intersection point $(\text{Re}_{24}, \lambda_{24})$

[4]For $\eta = 0.727$ its values are $\lambda_{24} \approx 2.9$ and $\text{Re}_{24} \approx 238 \approx 1.09\ \text{Re}_{cr}(0.727)$, and for $\eta = 0.5$ the values are $\lambda_{24} \approx 2.86$ and $\text{Re}_{24} = 74.2 \approx 1.09\ \text{Re}_{cr}(0.5)$. When normalized by $\text{Re}_{cr}(\eta)$, the neutral curves for different η, $0.5 \leq \eta < 1$, do not differ considerably, see also Figs. 1.9 and 1.11.

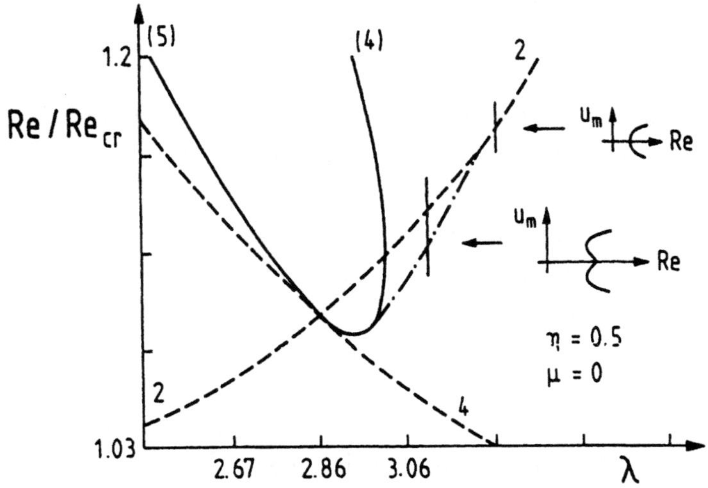

Figure 3.14: Andreichikov's analysis [An77, Fig. 4] in a neighborhood of the intersection point of the neutral curves for 2-vortex and 4-vortex flows.
- - - : Neutral curves for 2 and 4 vortices. —— : Curve of subharmonic bifurcation points marked (4) and (5) in Fig. 3.4. - · - : fold generated by subcritical bifurcation [MW2, Fig.1].

is a bifurcation point of codimension 2. Following John Bolstad, we call it the '(2, 4) **double point**' in the following.

When we studied the structure of families of solutions with respect to continuous variation of $\lambda = 2\pi/k\delta$ with fixed Re $= \sqrt{1.5}$ Re_{cr} and $\eta = 0.727$, $\mu = 0$, we found that the 2-vortex solution bifurcating from Couette flow on the low-λ side of the neutral curve (point (1) in Fig. 3.4) continuously changes with increasing λ into a 4-vortex solution which passes a fold point (6) at $\lambda \approx 2.68$ and then undergoes a period-halving bifurcation to its periodic repetition at point (5), $\lambda \approx 2.59$. When we started with maximum λ at the other side of the neutral curve in point (3), we found essentially the same thing, but without the fold point. The period-halving bifurcation occurs at $\lambda \approx 2.8$.

These phenomena are partially explained by the analysis of the solution set of eqs. (2.1)–(2.3) in a neighborhood of the point $(\mathrm{Re}_{24}, \lambda_{24})$ just defined. As already discussed in the section 'continuous change of period', Andreichikov [An75, An77] discussed qualitatively what can happen for arbitrary η and μ and gave a detailed analysis by numerical evaluation of the general formulas in the special case $\eta = 0.5$, $\mu = 0$. In Fig. 3.14 we repeat

Fig.4 of [An77], with adapted nomenclature, using the wavelength λ instead of a wave number α. Besides the neutral curves for one pair of vortices and for two pairs of vortices (or for n and $2n$ pairs in the n-th repetition), it gives the loci of the two subharmonic bifurcations in the (Re, λ)-plane. It also shows that there is a λ-interval adjacent to λ_{24} where λ-periodic Taylor vortex flows bifurcate subcritically from Couette flow. This generates a fold where the solution surface bends back to the supercritical domain. The size of this λ-interval is strongly η-dependent. While it is easy to detect this **subcritical fold** numerically for $\eta = 0.5$ (see the section 'continuous change of period'), more careful searching is necessary for $\eta = .727$. In the narrow gap limit $\eta \to 1, \quad \mu \to 1$ bifurcation from Couette flow is always supercritical (see section 'continuous change of period'). Computations by John Bolstad for $\eta, \ \mu$ approaching 1 suggest that the subcritical fold disappears by existing in a shorter and shorter λ interval, thus merging with the subcritical part of the curve of secondary bifurcations [Bol], which also disappears in the limit.

Andreichikov's analysis was confirmed by Tavener and Cliffe [TC] for $\eta = 0.615, \quad \mu = 0$ using Cliffe's code. These authors give much more details about the various stages of the interaction. Their figure of the neighborhood of the $(2,4)$ double point is repeated here as Fig. 3.30. It is shown later in this text since it includes neighborhoods of other double points as well. In Fig. 3.15 their analysis and additional information from a private communication [Tav] is summarized in a sketch of the solution surface in a neighborhood of $(\mathrm{Re}_{24}, \lambda_{24})$ for $\eta = 0.615$. For $\eta = 0.727$ this surface was computed by J. Huitfeldt [Hu92], as a test for newly developed numerical methods (see also [HR90]). It is shown here in Fig. 3.16.

The fold marked (6), (6′) in Fig. 3.4 seems to exist for all η commonly investigated, i.e. at least in the range $0.5 \le \eta < 1, \quad -0.2 \le \mu \le 0$: It has been detected for $\eta = 0.5$ [FM, Fig. 6], [RP86, Table I]; $\eta = 0.75$ [RP86]; $\eta = 0.95$ [Paf, Fig.1a] in the case $\mu = 0$, and also for $\eta = 0.5, \mu = -0.2$ [Jo82, Fig.2a].[5] Because of its existence for such a wide parameter range and because of its connection to the basic 2-vortex flow we call it the **basic** $(2,4)$-**fold**, following [DK] in part. Because of the periodic boundary conditions it has periodic repetitions. The copy of the fold on the $\times n$ branch is called the **basic** $(n, 2n)$-**fold**.

[5]The convergence problems reported in connection with the computations leading to [BO, Fig.6] in the case $\eta = 1, \quad \mu = 1$ should not be caused by this fold, see 'connections to the Rayleigh-Bénard problem'.

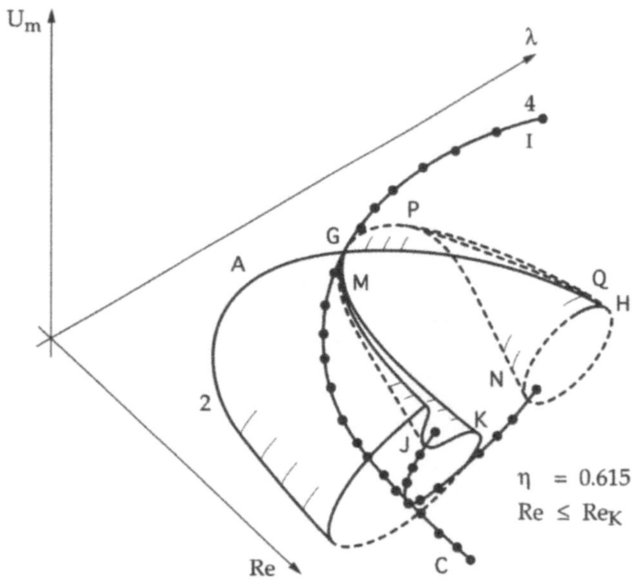

Figure 3.15: Sketch of the 2-vortex surfaces in a neighborhood of the $(2,4)$ double point for $\eta = 0.615$. This sketch summarizes the cross-sections given by Tavener and Cliffe [TC] and [Tav]. The capital letters are the same as used by these authors, see also Fig. 3.30; $G = (\mathrm{Re}_{24}, \lambda_{24}, 0)$. $(A)-(G)-(Q)-(H)$: On this curve the 2-vortex solutions bifurcate from Couette flow. $(C)-(G)-(I)$: Here the 4-vortex flows bifurcate from Couette flow. $(M)-(K)$: twin of the basic $(2,4)$-fold. $(G)-(J)$: Curve of $(\lambda \to 2\lambda)$-bifurcations on the upper branch of 4-vortex solutions (named (5) in Figs. 3.4 and 3.17). $(G)-(P)-(N)$: Curve of $(\lambda \to 2\lambda)$-bifurcations on the lower branch of 4-vortex solutions (named (4) in Fig. 3.17). $(P)-(Q)$ (2 lines): the 2 copies of the 'subcritical $(2,4)$-fold' generated by subcritical bifurcation of the 2-vortex solutions from Couette flow for wide gaps [SWM, Fig.1].

The basic $(2,4)$ fold is missing in Andreichikov's analysis. The reason seems to be that it does not exist in all neighborhoods of $(\mathrm{Re}_{24}, \lambda_{24})$, but emanates from the bifurcation curve (5) for a Reynolds number slightly larger than Re_{24}. For $\eta = 0.95$ M. Paffrath found it to separate from curve (5) for $\mathrm{Re} \approx 1.03\ \mathrm{Re}_{24}$ [Paf, Fig.1a]. Tavener/Cliffe (point 'M' in [TC, Fig. 2] and in Fig. 3.15) and M. Henderson [Hen] confirmed this observation for $\eta = 0.615$ and $\eta = 0.727$, respectively; see also Fig. 3.30.

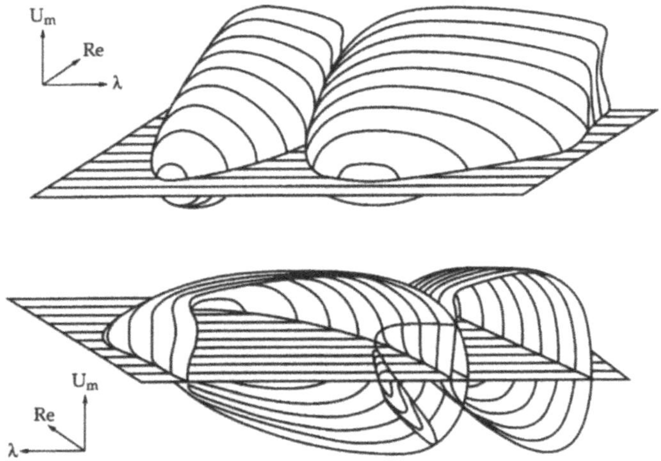

Figure 3.16: The 2-vortex surfaces and a periodic repetition in a neighborhood of the (2, 4) double point for $\eta = 0.727$ in north-west view (above) and south-west view (below) [Hu92, Fig. 4].

3.4.2 Connections to the Rayleigh-Bénard problem

As already described, there is a change of structure on both parts of the (2, 4)-surface, from a 2-vortex flow bifurcating from Couette flow to a 4-vortex flow undergoing a wavelength-halving bifurcation. This change of structure takes place continuously, thus indicating that the basic (2, 4)-fold originates from a perturbed bifurcation. As Busse pointed out [BO, Bus87], the unperturbed case cannot be found in the Taylor problem, but is provided by the limiting case $\eta \to 1$, $\mu \to 1$, for which equations (2.1)–(2.3) become mathematically equivalent to the Boussinesq approximation describing the Rayleigh-Bénard problem (see Chapters 1 and 4).

Thus we now consider convection layers between two infinitely extended parallel plates, with periodic boundary conditions in both the horizontal and vertical directions. The analysis of the bifurcation pattern in a neighborhood of the corresponding intersection of the neutral curves for 2-roll and 4-roll solutions yields, briefly (see [BO, Bus87, Bus3, Rie] and references therein):

There are a pure-mode 2-roll solution and its periodic repetition, a pure-mode 4-roll solution, and then two antisymmetric mixed-mode solutions called '**transition solutions**', and also non-antisymmetric mixed-mode

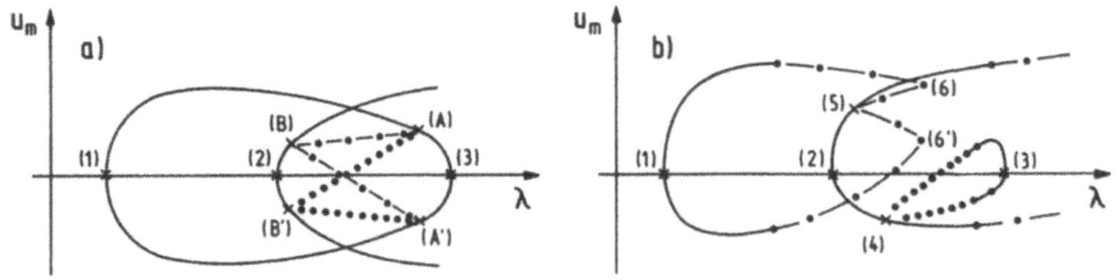

Figure 3.17: Breaking the midplane symmetry changes the bifurcation diagram. Radial velocity $u_m := u(1+\delta/2, 0)$ versus period λ.
a) Symmetric case (Rayleigh-Bénard problem). Isola (1) – (3) – (1): pure-mode 2-roll solutions bifurcating from the trivial solution in points (1) and (3) on the neutral curve. The upper and the lower branches contain $\lambda/2$-shifted twins. The periodic repetition of this loop bifurcates in point (2). Isola $(B)-(A)-(B')-(A')-(B)$: transition solutions. These are mixed-mode solutions with 2 large and 2 small rolls, interpolating between 2-roll and 4-roll solutions.
b) Unsymmetric case (Taylor problem). Bifurcation points $(A), (A')$ become perturbed and generate the fold points $(6), (6')$. Because of the lack of symmetry with respect to the midplane, 'inward transition solutions' (- · - , Fig. 3.27 c, d) and 'outward transition solutions' (· · ·, Fig. 3.28b) are clearly different [SWM, Fig.2].

solutions called '**mean component solutions**'. The two branches of transition solutions and their '$\lambda/2$-shifted twins' bifurcate from the 2-roll solution and from the 4-roll solution and thus connect them with a closed loop. On one of them the small rolls form where the flow goes inward. We call these solutions the **inward transition solutions.** (In [Jo81] they were called *alternating cell solutions*). On the other branch the small rolls form where the flow goes outward. We call these solutions the **outward transition solutions.** Their structures are similar to the ones given in the sketches of Fig. 3.6. The situation is very similar to the one described in [BKR, Fig. 4] : There are two branches of primary solutions, orthogonal to each other (the pure-mode solutions), and a closed loop of secondary solutions having growing and decaying components in both directions (the mixed-mode solutions). In Fig. 3.17a we sketch this situation.

Convection rolls are symmetric with respect to the midplane between the two plates, see eq. (4.68). This symmetry is broken in the Taylor problem because of the curvature of the two cylinders, and because of their different velocities. In Fig. 3.17b we sketch how the solution diagram is perturbed with changing η and μ. This figure also shows that, generically,

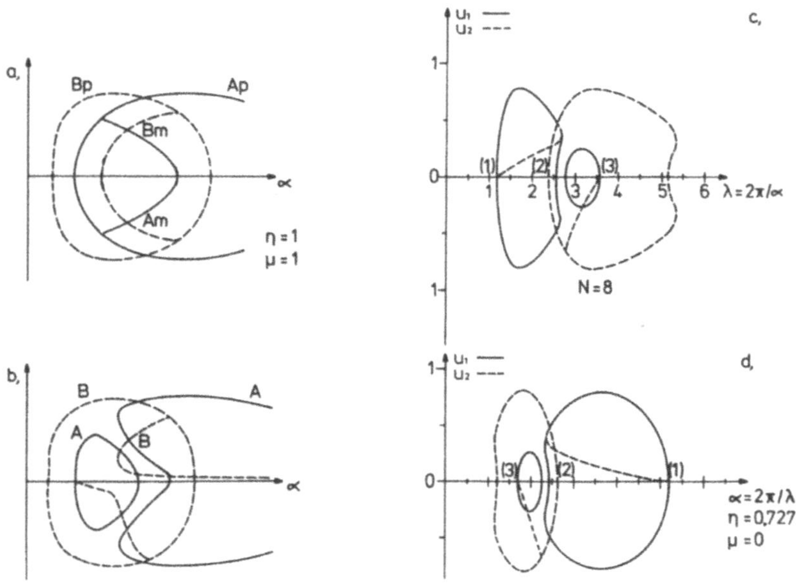

Figure 3.18: First (—) and second (- - -) Fourier components of the solution branches shown in Figs. 3.17 and 2.11 versus wave number α or wavelength λ, resp.
a) Rayleigh-Bénard problem, sketch, after Busse. The pure-mode solutions A_p (2-roll) and B_p (4-roll) bifurcate from the trivial solution. The mixed-mode solutions exist in the α-interval where A_m and B_m are both $\neq 0$. In the bifurcation points (B), (B') of Fig. 3.17 a $A_m = 0$ and $B_m = B_p$. In the bifurcation points (A), (A') $B_m = 0$ and $A_m = A_p$.
b) Perturbed Rayleigh-Bénard problem, after Busse.
c) Taylor problem. First two Fourier components at the midpoint $(r_m, 0)$ versus λ, computed with $N = 8$, $M = 31$.
d) Taylor problem. Same as part c), but versus wave number $\alpha = 2\pi/\lambda$. The similarity between *b*) and *d*) is clearly seen. Since our computations treat a wide-gap problem, the perturbations are quite large [SWM, Fig.3].

the change of structure of the flows does not occur at the fold point, but somewhere else. This was observed in all our computations.

In Fig. 3.18 we demonstrate the perturbation in a different projection. Part *a* (Rayleigh-Bénard problem) and part *b* (perturbed Rayleigh-Bénard problem) are corrected versions of [Bus87, Figs. 1a, 1b], while parts *c* and *d* result from computations of the Taylor problem. This shows that the asymmetries of the solution branches about the λ-axis shown in Figs. 3.17b and 2.11a are mainly caused by the perturbation of the second Fourier component.

Why do the bifurcations on the 2-roll branch become perturbed, while those on the 4-roll branch persist? The answer can be found from symmetry considerations as presented in [TC, Wer] : In the Rayleigh-Bénard problem, all four bifurcations $(A), (A'), (B), (B')$ are symmetry-breaking, and thus of the pitchfork type. In the Taylor problem, the symmetry of the 2-vortex solution is lost because of the asymmetries of the apparatus. The 2-vortex solution is invariant with respect to the identity operator only. (The mirror symmetry with repect to $z \equiv 0$ is a property of all our solutions and thus does not count here.) The codimension-zero singularity is thus a quadratic fold point – bifurcations $(A), (A')$ turn into the fold points $(6), (6')$. The 4-vortex solution still is invariant with respect to translation by $\lambda/2$. The loss of this invariance is equivalent to loss of reflectional symmetry at a distance $\lambda/4$ from the midplane $z \equiv 0$. The codimension-zero singularity is thus a pitchfork symmetry-breaking bifurcation [TC]: Bifurcation points $(B), (B')$ turn into bifurcation points (5) and (4). Note the distortion in the parameter space: $(B), (B')$ occur at the same λ-value, but (5) and (4) do not.

3.4.3 The basic $(n, 2n)$-fold for higher Reynolds numbers

Varying λ with fixed $\mathrm{Re} = \sqrt{1.5}\,\mathrm{Re}_{cr}$ (or $T = 1.5\,T_{cr}$), we found a fold point near $\lambda = 2.68$ and a period-halving (or period doubling) bifurcation near $\lambda = 2.59$. Varying the Taylor number T or Re^2 with fixed $\lambda = 2.65$, we also found a fold point and a period-halving bifurcation, as shown in Fig. 3.8. By a following investigation [DK], both fold points were shown to belong to the same fold, the basic $(2, 4)$-fold ($\times 1$ version of the basic $(n, 2n)$-fold). In [DK] it was computed with a preliminary fold-following algorithm up to $\mathrm{Re}^\delta \approx 290$ or $\mathrm{Re} \approx 3.53\,\mathrm{Re}_{cr}$ ($T \approx 12.3\,T_{cr}$), and then convergence problems of the multi-grid iterations made it impossible to continue. These could be due to one of the weaknesses of that code, to the fact that the basic grid was chosen too coarse, and/or to the fact that there is another fold nearby: we obtained Fig. 3.21 for the value $\mathrm{Re} = 3.65\,\mathrm{Re}_{cr}$, which is quite close to the value $\mathrm{Re} \approx 3.53\,\mathrm{Re}_{cr}$. Using pseudo-arclength continuation with respect to λ for a number of Re values, and pseudo-arclength continuation with respect to Re for two fixed λ values, we [MW2] found that this fold still exists at $T = 21T_{cr}$. But its locus proved to be very sensitive to discretization errors, and the machine first used (CRAY 1 of ZIB Berlin) was not big

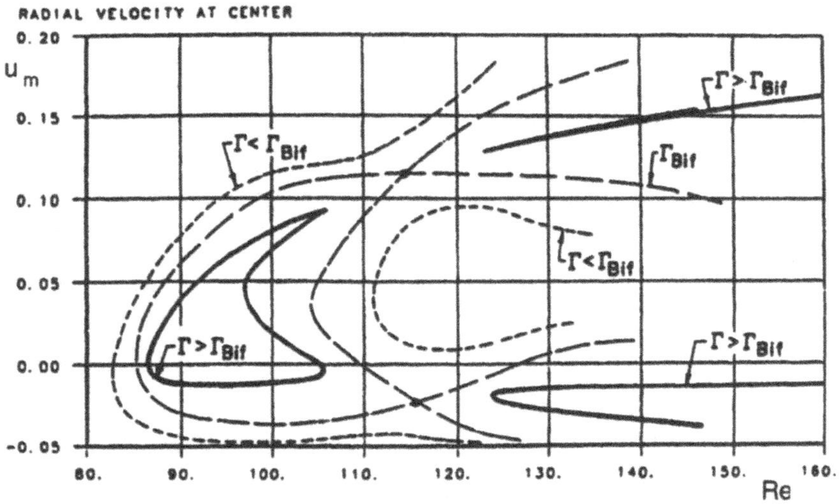

Figure 3.19: Sketch of possible bifurcation in the $(\mathrm{Re}^\delta, u_m)$-plane, $\eta = 0.727$, $\mu = 0$. Solid curves are computed paths for $\lambda \equiv \Gamma = 2.65$ $(\lambda > \lambda_{bif})$, as in Fig. 3.8. Dashed curves are qualitative, based on incomplete calculations. Short dashes are for $\lambda < \lambda_{bif}$, long dashes for $\lambda = \lambda_{bif}$ [DK, Fig. 9].

enough to get satisfactory results. The solutions for $\mathrm{Re} = 3.65\ \mathrm{Re}_{cr}$ labeled '5' and '93' in Fig. 3.10 and in Fig. 3.21 are very similar to each other. But they seem to belong not only to different branches, but even to different connected components of the solution set. There might, however, be a point in parameter space where these components meet in a bifurcation point: we repeat the speculation of Dinar/Keller about such a bifurcation point as Fig. 3.19. To our knowledge, it still has not been confirmed or disproved. In Fig. 3.20 we display the structure of both solutions, and in Fig. 3.21 we show portions of both branches in two different projections. Computations by John Bolstad with a proper fold-following algorithm detected several cusps of the basic (2, 4) fold [Bol92, Fig. 2]. For $\eta = 0.727$, the first such cusp was found at $\mathrm{Re} \approx 3.554\ \mathrm{Re}_{cr}$, $\lambda \approx 4.165$, see Fig. 4.14 [Bol][6].

Our experience is that the change of structure along a branch usually does not occur at the fold point, but somewhere else. The exact locus in parameter space is usually affected by discretization errors. This includes: If the change of structure occurs near the fold point, then the side of the fold

[6]Computations along the four sides of a rectangle surrounding that point verified that it is a cusp point *for the numerical approach of that code.*

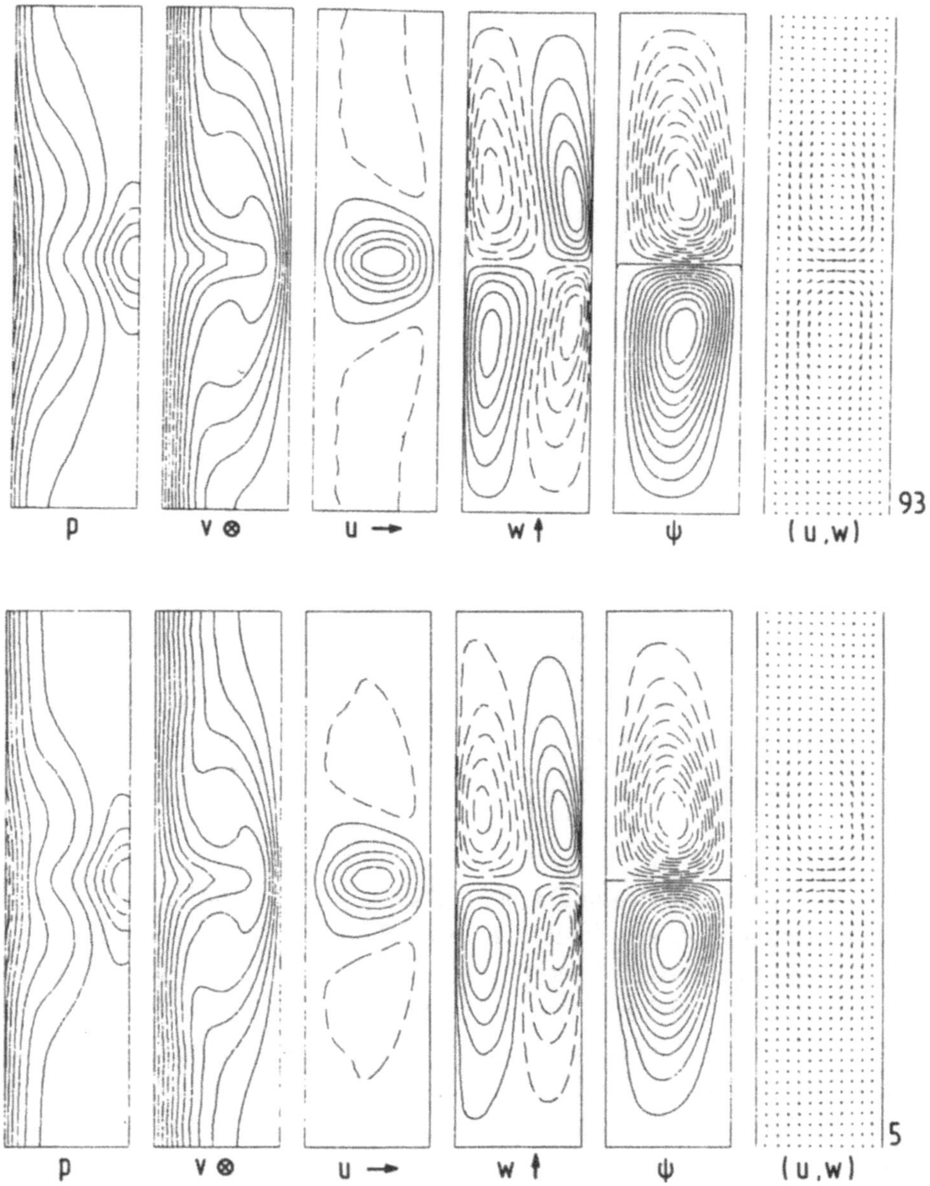

Figure 3.20: Level lines of reduced pressure p, velocity components v, u, w, stream function ψ, and (u, w) velocity field in the (r, z)-plane for the two solutions marked '93' ($\lambda = 3.9008$) and '5' ($\lambda = 4.1539$) in Figs. 3.9, 3.11 and 3.21; Re $= 800 \approx 3.65$ Re_{cr}, $\eta = 0.727$, $\mu = 0$; M $= 31$, N $= 12$ [MW2, Fig.2].

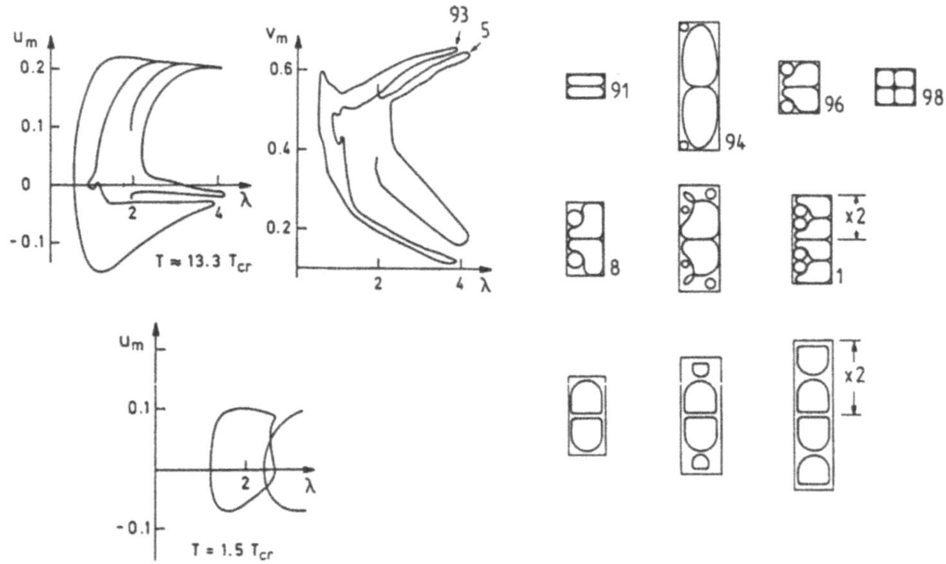

Figure 3.21: Radial velocity u_m versus λ and azimuthal velocity v_m versus λ for parts of two $T = 13.3T_{cr}$ branches and for the $T = 1.5T_{cr}$ branch of Fig. 3.4; $\eta = 0.727$, $\mu = 0$. Note that both u_m versus λ plots are drawn to the same scale. The change of flow pattern on these branches is only sketched here. Detailed pictures of the solutions on the $T = 13.3T_{cr}$ branches are given in Figs. 3.10, 3.12, and 3.20. M = 31, N = 6 for $T = 1.5T_{cr}$ and M = 31, N = 12 for $T = 13.3T_{cr}$ [MW2, Fig.3].

point on which it occurs might depend on the discretization used. For the basic $(n, 2n)$-fold we also found the locus of the fold to be strongly affected by discretization errors of the Fourier approximation. We discuss the role of Fourier discretization errors in the computation of this fold near Figs. 2.9b and 2.10, which are closely related to the upper part of Fig. 3.21. As is also shown in Chapter 2, all studies of the discretization error of the code Tayperio [MK80, FM, MK85, SWM] have confirmed that the discretization error affects only the details, but not the main structures observed, as long as enough Fourier components are used so that solutions do not completely disappear. As is pointed out in various places in this text, good agreement was found with the results obtained with other codes [BoKe, TC, RP86], though these codes use different discretization methods[7]. The discretization

[7]These methods are: multi-grid [BoKe], finite elements [TC], and Galerkin method with Fourier modes in the axial and Čebyšev polynomials in the radial direction [RP86].

errors are thus of a different nature, and the unfoldings caused by numerical approximation are likely to be different.

The main effect of the discretization errors in the code Tayperio is a shift of whole structures in parameter space, and possibly also a deformation of these structures. This is discussed in more detail in Chapter 2; see also Fig. 4.8.

3.4.4 The basic 2-vortex surface

As has just been pointed out, we still cannot make a final statement about the basic $(n, 2n)$-fold for higher Reynolds numbers. The form of the fold, its locus in parameter space, and maybe even its existence for higher Re are subject to discretization errors. The existence of the two surfaces meeting in the fold, however, and also the structure of the solutions on these surfaces, do not seem to be affected by discretization. In Figs. 2.9b and 2.10 we showed a cut through the surface for Re $= 800 \approx 3.65$ Re_{cr} in two different projections as computed with three different discretizations errors. For all three discretizations we found the same patterns: bifurcation of a 2-vortex flow from Couette flow on the neutral curve; growth of both vortices with a growing difference between the maximum outward and the maximum inward speed; generation of two additional vortices in the inward region and thus formation of an inward transition solution; formation of a double-vortex flow; bifurcation to Couette flow on the eigencurve generated by the second eigenvalue of the linear differential operator related to Couette flow.

The double-vortex flow was not found for Re $= \sqrt{1.5}$ Re_{cr}; see Fig. 3.21. Thus we now investigate how the surfaces of these two cuts for fixed Re are connected for fixed λ and varying Re. Figure 3.22a shows non-interacting branches for $\lambda = 1$ and $\lambda = 2$. Figure 3.22b shows three solutions on the $\lambda = 2$ branch connecting the inward transition solution on the Re $= 800$ branch of Fig. 3.9 with the 2-vortex solution for $\lambda = 1$ in a subharmonic bifurcation. Figure 3.23 gives the same type of investigation for λ-values close to the 'nose' in Fig. 3.9a. For $\lambda = 1.2$ there is still subharmonic bifurcation. For $\lambda = 1.1$, however, there is bifurcation of the double-vortex flow to Couette flow. Figure 3.24 gives the bifurcation curve of the double-vortex flow from Couette flow together with the basic neutral curve, and periodic repetitions of both curves. Note that both curves intersect with the first periodic repetition of the basic neutral curve. These two intersection points $(\mathrm{Re}_{24}, \lambda_{24})$ and $(\mathrm{Re}_{d4}, \lambda_{d4})$ seem to be connected by a curve of subharmonic bifurcations: period-halving bifurcations between

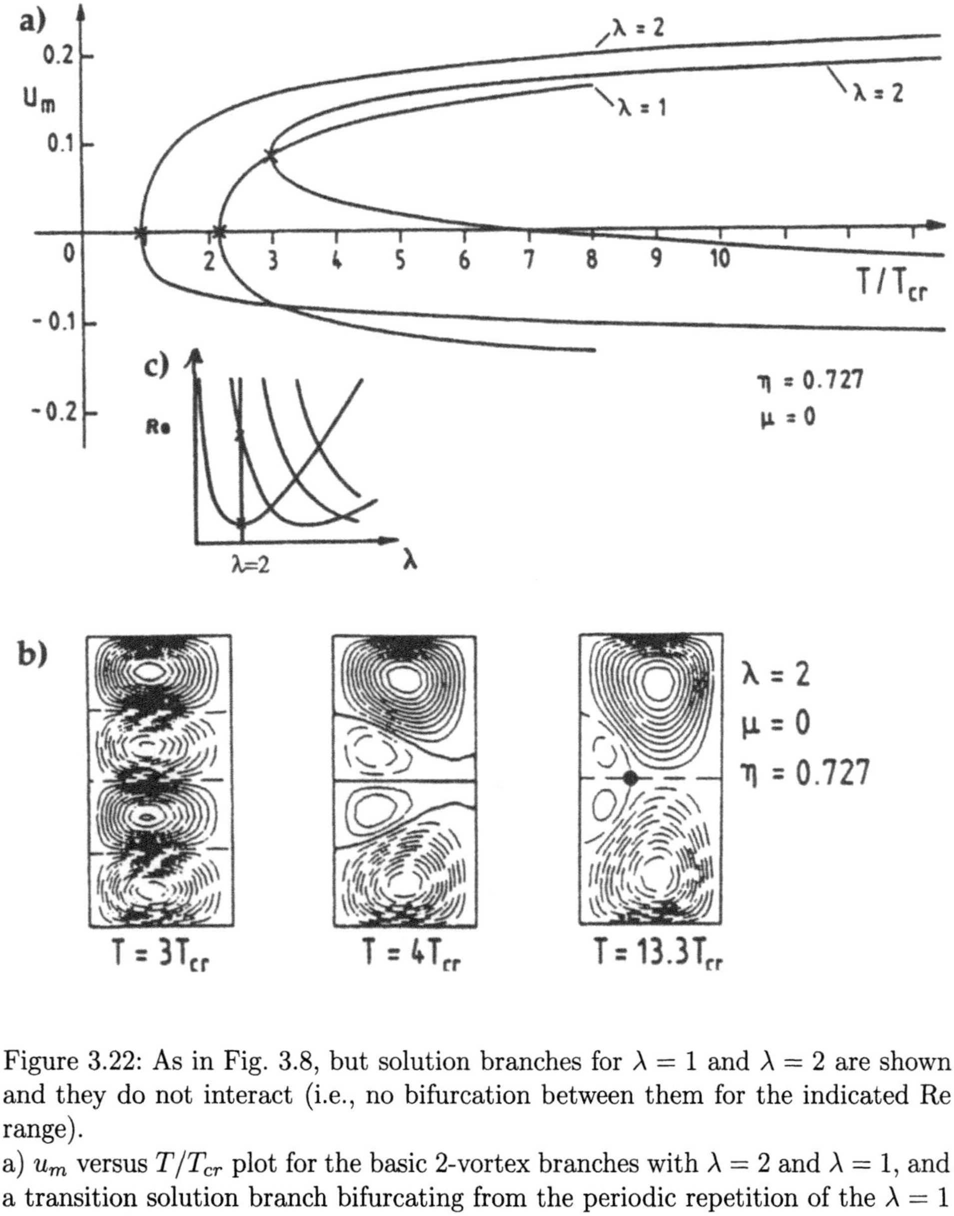

Figure 3.22: As in Fig. 3.8, but solution branches for $\lambda = 1$ and $\lambda = 2$ are shown and they do not interact (i.e., no bifurcation between them for the indicated Re range).
a) u_m versus T/T_{cr} plot for the basic 2-vortex branches with $\lambda = 2$ and $\lambda = 1$, and a transition solution branch bifurcating from the periodic repetition of the $\lambda = 1$ branch. Bifurcation points are indicated by "×". The other intersection points are due to the projection used.
b) Level lines of the stream function ψ for the three mixed-mode solutions on the branch shown in part a. •: free stagnation point. Computations with $M = 31$, $N = 8$ [MW2, Fig.7].

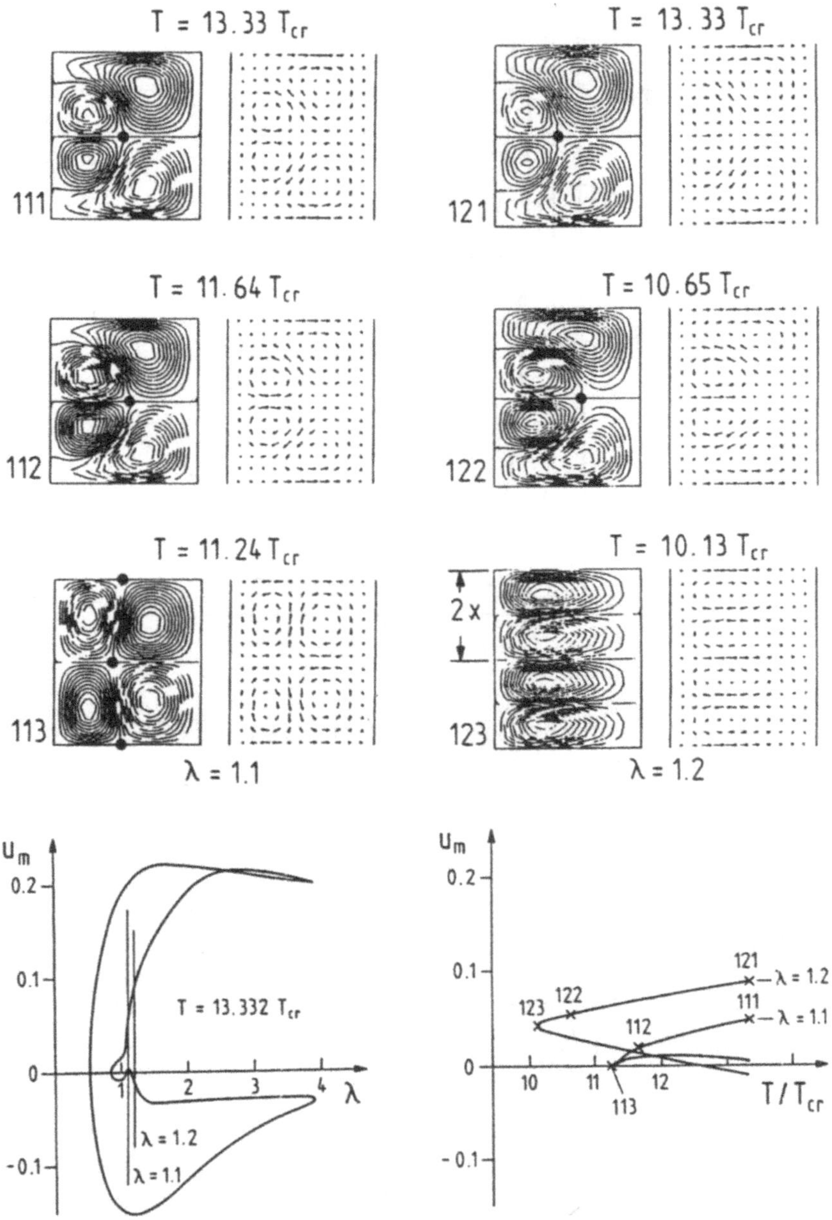

Figure 3.23: Further investigation of the surface. While there is still subharmonic bifurcation for $\lambda = 1.2$, there is a continuous change into a double-vortex flow and bifurcation to Couette flow for $\lambda = 1.1$. •: free stagnation point. $M = 31,\ N = 12$ [MW2, Fig.8].

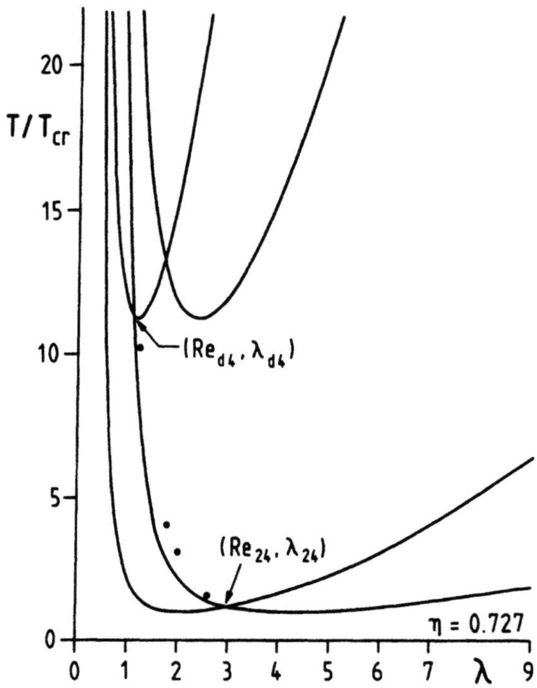

Figure 3.24: Neutral curve for Couette flow (first eigenvalue) and bifurcation curve of the double-vortex flow (second eigenvalue) in the $(\lambda, T/T_{cr})$-plane, together with their first periodic repetitions. •: subharmonic bifurcation of type (5) as described in the text [MW2, Fig.9].

the inward transition solution with spatial period λ and the basic 2-vortex flow with spatial period $\lambda/2$. Until now, we have seen four such bifurcation points: For $T = 1.5T_{cr}$ (point (5) of Fig. 3.4), for $T = 4T_{cr}$ (Fig. 3.26a), for $\lambda = 2$ (Fig. 3.22), and for $\lambda = 1.2$ (Fig. 3.23). They are also shown in Fig. 3.24.

The conjecture that the two double points are connected by the curve of points (5) was later checked both with a bifurcation-following algorithm for the Taylor problem and by calculations for a model problem derived from the Rayleigh-Bénard problem. The results are dealt with in the next chapter. It would also be very interesting to see an analysis of the bifurcation pattern in a neighborhood of the double bifurcation point $(\mathrm{Re}_{d4}, \lambda_{d4})$. Does the other curve of subharmonic bifurcations ((4) in Fig. 3.4) also go through this point? From the analysis related to Fig. 3.18 we expect that this will happen. This will also be discussed further in the next chapter.

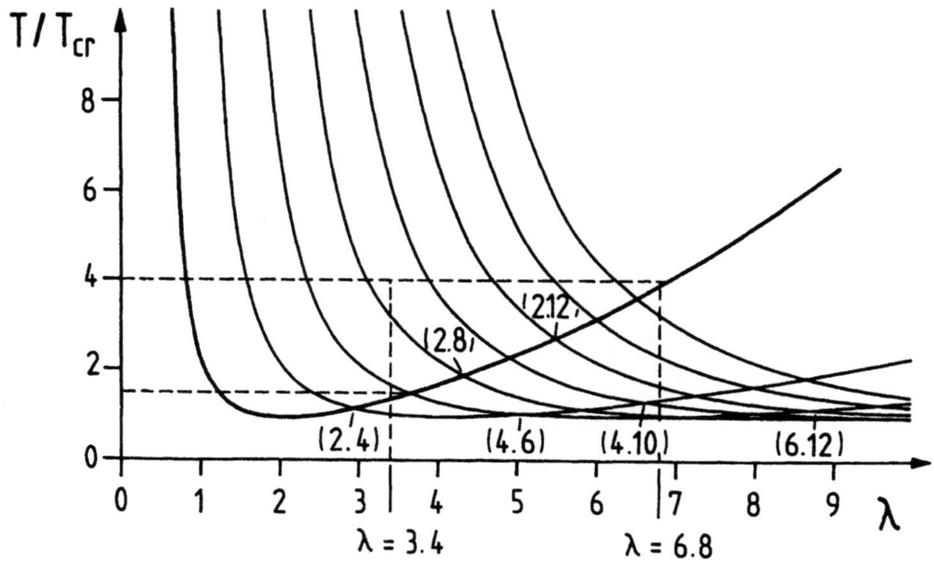

Figure 3.25: Neutral curve of 2-vortex flows with those of its periodic repetitions which play a role in our $\mathrm{Re} = 2\ \mathrm{Re}_{cr}$ computations [SWM, Fig.4].

If we now complete Fig. 3.24 in our mind by all the periodic repetitions of the curves given there and if we imagine that there will also be curves of higher-order subharmonic bifurcations and folds, and they all intersect and generate new curves of bifurcation points and folds where they intersect .... Then it's not surprising that there are so many different solutions, and branches as complicated as the one displayed in Fig. 3.11.

3.5 $\mathrm{Re} = 2\ \mathrm{Re}_{cr}$ and the (n, pn) double points

In the previous section we described how the solution diagrams for relatively small λ-values are influenced by singularities emanating from the double points $(\mathrm{Re}_{24}, \lambda_{24})$ and $(\mathrm{Re}_{d4}, \lambda_{d4})$. Here we describe a solution diagram for $\mathrm{Re} = 2\ \mathrm{Re}_{cr}$ and varying λ in the whole λ-range inside the neutral curve for Couette flow. From Fig. 3.25 it can be expected that the double points $(\mathrm{Re}_{2,q}, \lambda_{2,q}), q = 4, 6, \ldots, 16$ play a role, i.e. the (n, pn) double points, $n = 2,\ \ 2 \leq p \leq 8$, and possibly even the $(m, m + p)$ double points, $m = 4, 6, 8;\ \ m + p \leq 16$. The computed solution diagrams are presented in Fig. 3.26, in four different projections.

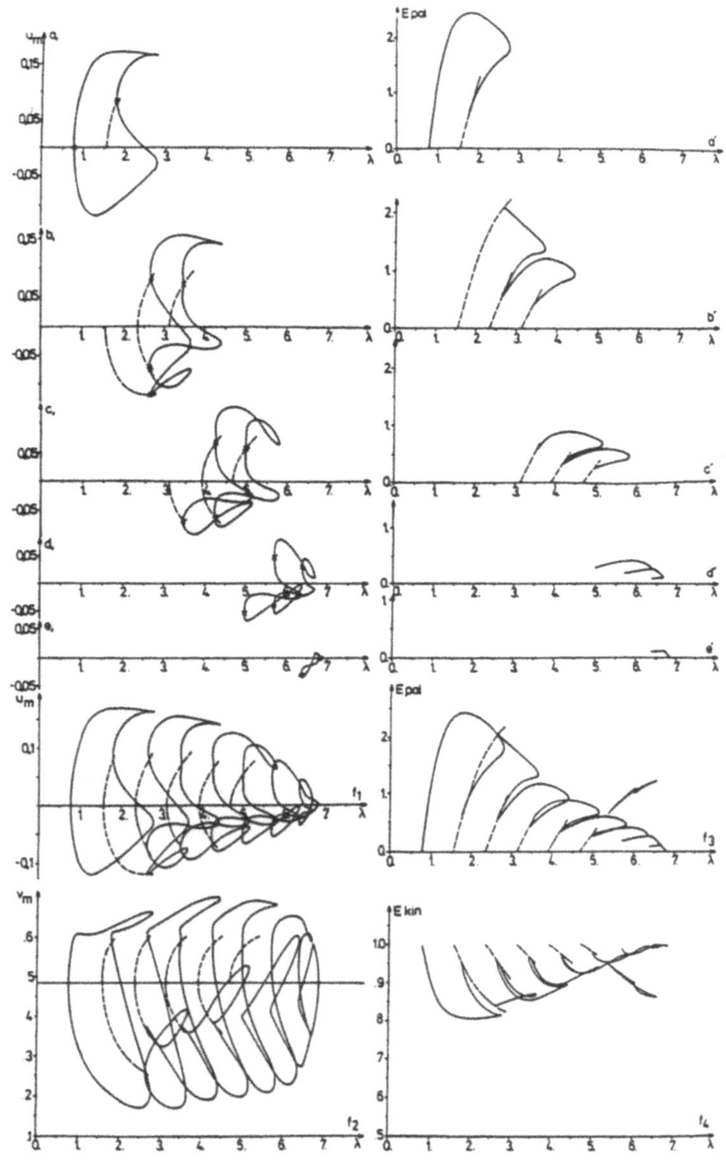

Figure 3.26: Solution diagrams for Re $= 2$ Re_{cr}.
— : investigated solution branches. – – : periodic repetitions of the basic branch. $a) - e), f_1)$: Radial velocity u_m versus period λ. $f_2)$: Azimuthal velocity v_m vs. λ. $a') - e'), \; f_3)$: Poloidal kinetic energy E_{pol} vs. λ. Same branches as in $a - e, \; f_1$. $f_4)$: Total kinetic energy E_{kin} vs. λ. $f_3)$ and $f_4)$ show also the energies of the branches displayed in Fig. 3.33 [SWM, Fig.5].

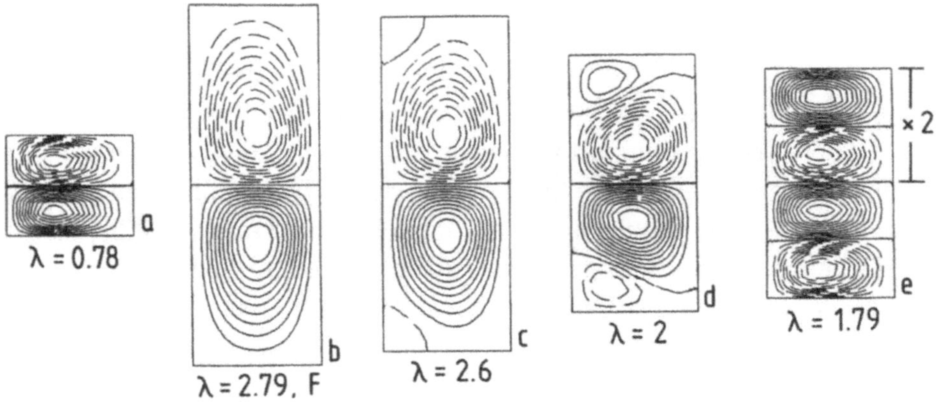

Figure 3.27: Level lines of the stream function ψ for some of the flows on the branch shown in Fig. 3.26a. The branch has a fold point for $\lambda \approx 2.79$ (basic (2, 4)-fold). Computations with $N = 8, M = 31$. The numerical error ε defined in eq. (2.115) satisfied $10^{-5} \leq \varepsilon \leq 10^{-2}$, and at least the last 4 coefficients were monotonically decreasing [SWM, Fig.6].

Figure 3.26a shows a familiar picture, very similar to the isola (1) – (6) – (5) – (6′) – (1) of Figs. 3.17b and 2.11a: Bifurcation of a 2-vortex flow from Couette flow, fold point, continuous change into a 4-vortex flow, and wavelength-halving bifurcation to the periodic repetition (×2-branch). Pictures of these flows are given in Fig. 3.27, and in more detail in [WS87, Figs.43-48] ; see also Fig. 3.6 or Fig. 3.22 for similar flows. Because of the periodic boundary conditions in z , this isola is periodically repeated, as discussed earlier and as shown in Fig. 3.7 for $T = 1.5T_{cr}$. For the case $T = 4T_{cr}$ considered here, the growth of the number of solutions for increasing λ is even more dramatic. This is because the fold is more pronounced here. There is thus more overlapping of the various periodic repetitions. Besides this, we found additional solution branches connecting these periodic repetitions. They are shown in the different sections of Fig. 3.26, together with those parts of the periodic repetitions which are involved in bifurcations.

We now describe the isola given in Fig. 3.26b. It connects the ×2, ×3, ×4-branches. The corresponding flow pictures are given in Fig. 3.28a - n, and in more detail in [WS87, Figs.50-65], where the level lines of pressure and velocity components are also given. Let us look at the bifurcations which occur on this isola: The bifurcation on the ×2-branch is of type (4), Fig. 3.17b. It is a secondary bifurcation induced by the (2,4) double point.

The solutions between this bifurcation point and the fold point given in Fig. 3.28b are 'outward transition solutions' defined near Fig. 3.17. They lie on the surface bifurcating from Couette flow at the eigenvalue curve of the 2-vortex flow, as can be seen from Fig. 3.29 (solutions ζ, ζ').

There are two bifurcations on the $\times 3$-branch. These are the secondary bifurcations induced by the $(2,6)$ double point. This can be seen by continuing in decreasing Reynolds number for solutions β and ε of 3.29: Solution ε lies between the solutions shown in Fig. 3.28b and 3.28c, while solution β lies between the solutions of Figs. 3.28h and 3.28i. The flow patterns we obtained by this continuation were so similar to those shown in Figs. 3.28 c - i that we omit them here.

The neighborhood of the $(2,6)$ double point (intersection of the $\times 1$ with the $\times 3$-branch) was investigated in detail by Tavener/Cliffe [TC, Figs. 6 and 7] and by Bolstad [Bol93], see also Fig. 3.30. Our results are in good agreement with theirs for those parameter values for which results were obtained with both codes. We independently also found:

i) The secondary bifurcations from the $\times 3$-branch are transcritical, and they both occur at the same λ-value. The deviation from a pitchfork bifurcation is very small (bifurcation at $\lambda \approx 2.65$, fold point at $\lambda \approx 2.63$). It is hardly visible in both Figs. 3.29a and b. The flow at one of the corresponding fold points for $\mathrm{Re} = 2\ \mathrm{Re}_{cr}$ is shown in Fig. 3.28f. Both secondary bifurcations occur at exactly the same λ-value since they are $\lambda/2$-shifted twins.

ii) There is another fold (in two copies) emanating from the $(2,6)$ double point. We shall call it *the basic* $(2,6)$*-fold*, or, more generally, **the basic** $(n, 3n)$**-fold**. The streamlines of one of the flows at the fold points for $\mathrm{Re} = 2\ \mathrm{Re}_{cr}$ are given in Fig. 3.28j.

It seems reasonable to assume that the $(2,6)$-interaction in the Rayleigh-Bénard problem produces four secondary pitchfork bifurcations (see Chap. 4, near Fig. 4.10). Very similar to the case sketched in Fig. 3.17a, but this time with the $\lambda/2$-shifted twins meeting on the 2-vortex branch. In the Taylor problem, these two bifurcations become perturbed again, and they generate the two copies of the basic $(2,6)$-fold. The secondary bifurcations on the $\times 3$-branch become transcritical in the perturbation.

But this still does not explain the folds encountered between ζ and ε (or ζ' and ε', respectively) in Fig. 3.29b. In the sketch in Fig. 3.31 we suggest that they are generated by perturbed tertiary bifurcations between the mixed-mode solutions of the $(2,4)$- and the $(2,6)$-double point. Proceeding

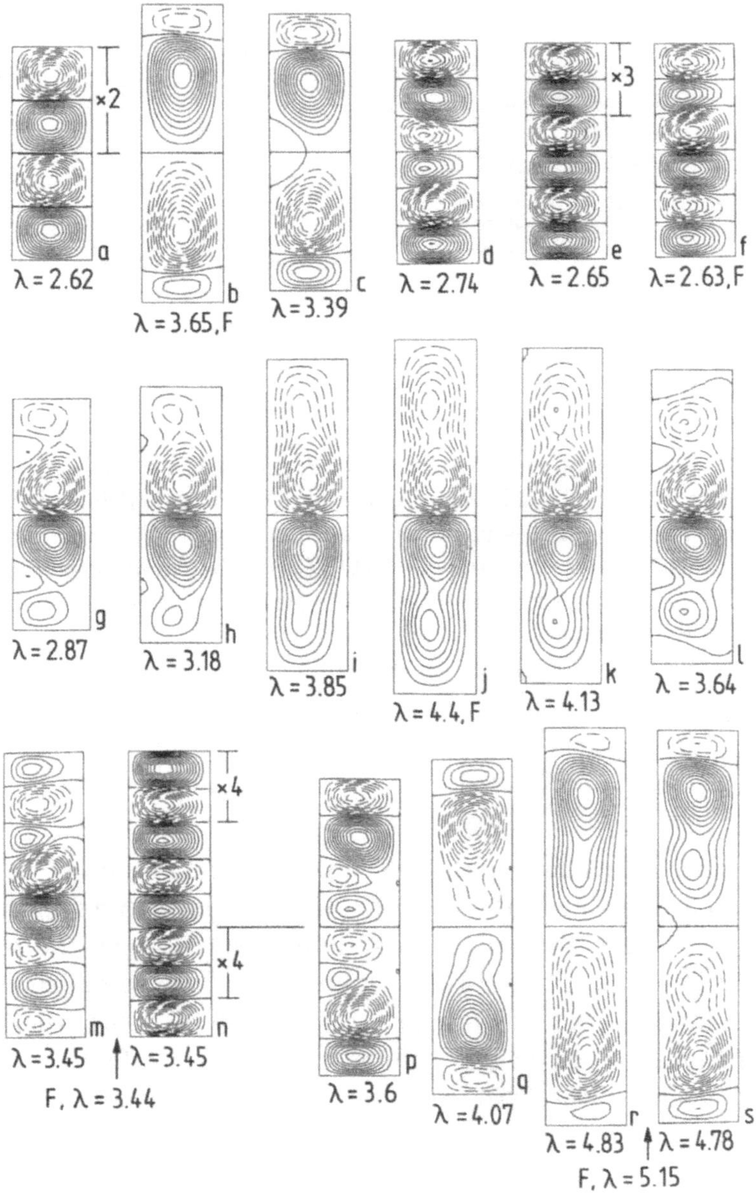

Figure 3.28: Level lines of the stream function ψ for some of the flows on the branches shown in Fig. 3.26b, b' (parts a–n) and in Fig. 3.26c (parts p–s). The branch of the flows shown in parts p–s bifurcates from the $\lambda/8$-shifted twin of the flow shown in part n. F: fold point [SWM, Fig.7].

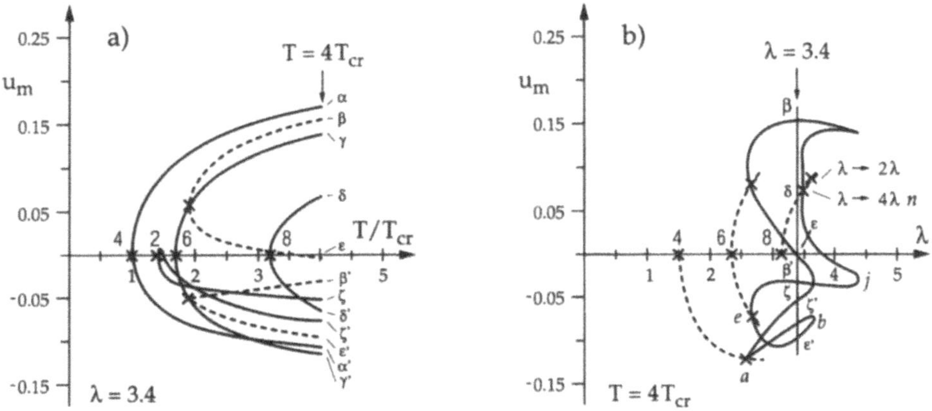

Figure 3.29: Analysis of the solution set for λ near 3.4, $T \leq 4T_{cr}$, $\mathrm{Re} \leq 2\ \mathrm{Re}_{cr}$. Greek names refer to the solutions for $T = 4T_{cr}$, $\lambda = 3.4$. Primes indicate $\lambda/2$-shifted twins. Some branches meet their twins, others do not.
a) —— : branches bifurcating from Couette flow on the curves shown in Fig. 3.25 for $\lambda = 3.4$. - - - - : secondary branches generated by the $(2,6)$ double point. $N = 12$, $M = 31$, error $\varepsilon < 10^{-1}$.
b) Enlarged version of Fig. 3.26b. Those solutions lying on the periodic repetitions of the basic branch are omitted. Roman names refer to flows shown in Fig. 3.28. $N = 14$, $M = 31$, $10^{-4} \leq \varepsilon < 10^{-1}$, last 8 coefficients monotonically decreasing [SWM, Fig.8].

along this line, and assuming that all (n, pn) interactions, $n = 2$, $2 \leq p \leq 8$, are essentially the same in the Rayleigh-Bénard problem, we could also draw a sketch showing a similar nonlinear interaction between the other two secondary branches of the $\times 3$-branch and two of the secondary branches of the $\times 4$-branch. On the $\times 4$-branch, the $\lambda/2$-shifted twins meet. The two secondary bifurcations on the $\times 4$-branch are thus not forced to occur at the same λ-value in the Taylor problem. We found them at $\lambda = 3.4535$ and at $\lambda = 3.4583$, respectively.

There are two fold points close to the secondary bifurcation point shown in Figs. 3.29b and 3.28n. These two folds could be generated by a nonlinear interaction of the two secondary bifurcations on the $\times 4$-branch shown in Fig. 3.29b for $\mathrm{Re} = 2\ \mathrm{Re}_{cr} = 438.2$: One of them is generated by the $(2,8)$ double point and belongs to the isola just discussed. Figure 3.29b shows the branch bifurcating there. The other secondary bifurcation is generated by the $(4,8)$ double point, the periodic repetition of the $(2,4)$

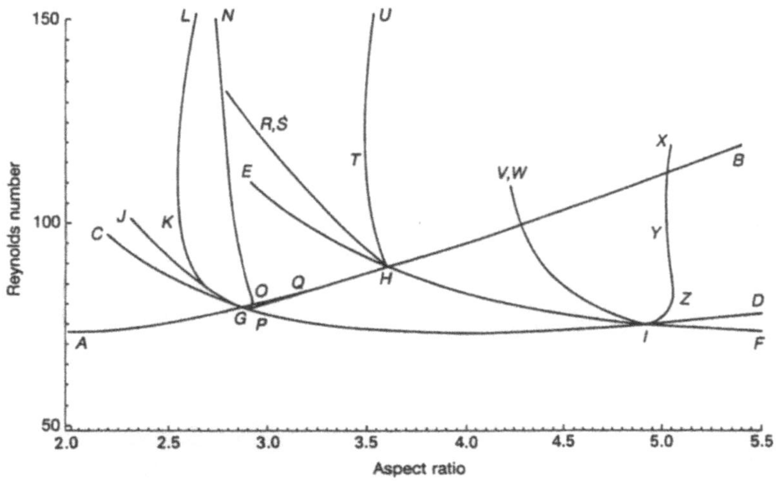

Figure 3.30: Loci of bifurcations and folds in the (λ, Re)-plane in a neighborhood of the $(2,4)$, $(2,6)$ and $(4,6)$ double points for $\eta = 0.615$ [TC, Fig. 10]. In addition to these, Bolstad [Bol93, Fig. 1] found a curve of fold points emenating from the point Z.

double point. Figure 3.29b only shows that point. The branch bifurcating there is the periodic repetition of the branch shown in Fig. 3.26a. For $\lambda =$ 3.4 these two secondary bifurcations both occur between Re $=$ 445. and Re $=$ 466. They are thus not shown in Fig. 3.29a.

An inspection of Figs. 3.26c and d shows that the isolas connecting the $\times 4, \times 5, \times 6$-branches and the $\times 6, \times 7, \times 8$ branches, respectively, are similar to the isola of Fig. 3.26b. The loop shown in Fig. 3.26e then 'gathers the left-overs'. The similarity between the $n, (n+1), (n+2)$-transitions can also be seen from the flow pictures given in [WS87, Abb. 66-109]. But things become even more involved. Too many perturbed bifurcations seem to be present. We thus confine ourselves to showing a small part of the $\times 4, \times 5, \times 6$-isola (Figs. 3.28 p - s), and to mentioning general, repeated features.

Figure 3.32 shows what the flows corresponding to the one shown in Fig. 3.28b look like in the higher transitions. There are similar series (Figs. 3.28c, 3.28s, Abb. 94), (Figs. 3.27b, 3.28j, Abb. 83), and (Figs. 3.27d, 3.28l, Abb. 84, Abb. 105). Here 'Fig.' refers to this text, while 'Abb.' refers to figures of [WS87]. The feature common to these series is: The two large vortices of these flows always contain an increasing number of small vortices

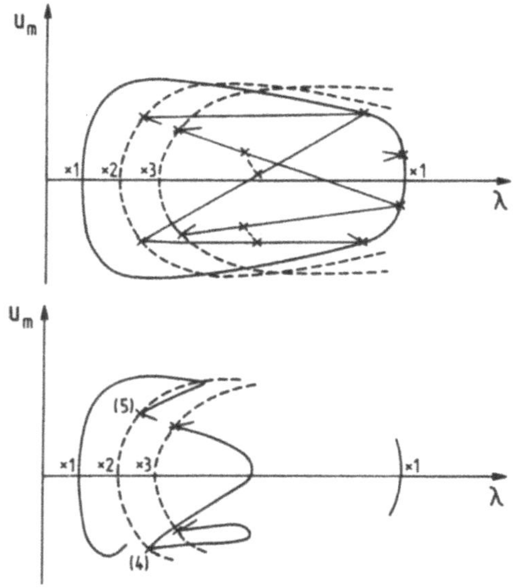

Figure 3.31: Attempt to explain the interaction between the secondary branches of the $(2,4)$ and the $(2,6)$ double points. *a)* Possibly existing branches in the Rayleigh-Bénard problem. *b)* Perturbed case, Taylor problem. Details are given in the text [SWM, Fig.9].

inside. These are weak for part of the branch, e.g. in Figs. 3.28i, 3.28 q-r. And then this structure breaks up to allow many vortices, as in Figs. 3.28 l, m, n or Figs. 3.28 q, p.

To our knowledge, Taylor vortex flows with a separatrix (i.e. co-rotating small vortices separated by a stagnation point inside large vortices), were first reported in the periodic boundary case in [FM, Fig. 8] (repeated here as Fig. 3.3), in connection with computations near the $(2,6)$ double point[8]. In the finite cylinder case and in the Rayleigh-Bénard problem they have been computed earlier, see [BH79] and the references therein. In the time-dependent Rayleigh-Bénard problem, solutions with such a structure were also computed [Nev87, Fig. 10, $t = 2000$]. Note that these flows are similar to flows between rotating spheres. Compare, for instance, the $\lambda/2$-shifted version of Fig. 3.28 q, r with some of the flows shown in [Scr, Figs.11-13]. And there is also formal similarity of the experimentally observed (!)

[8]Near Fig. 4.13 we show that such flows can be produced by linear superposition of the modes dominating the 2-6 interaction.

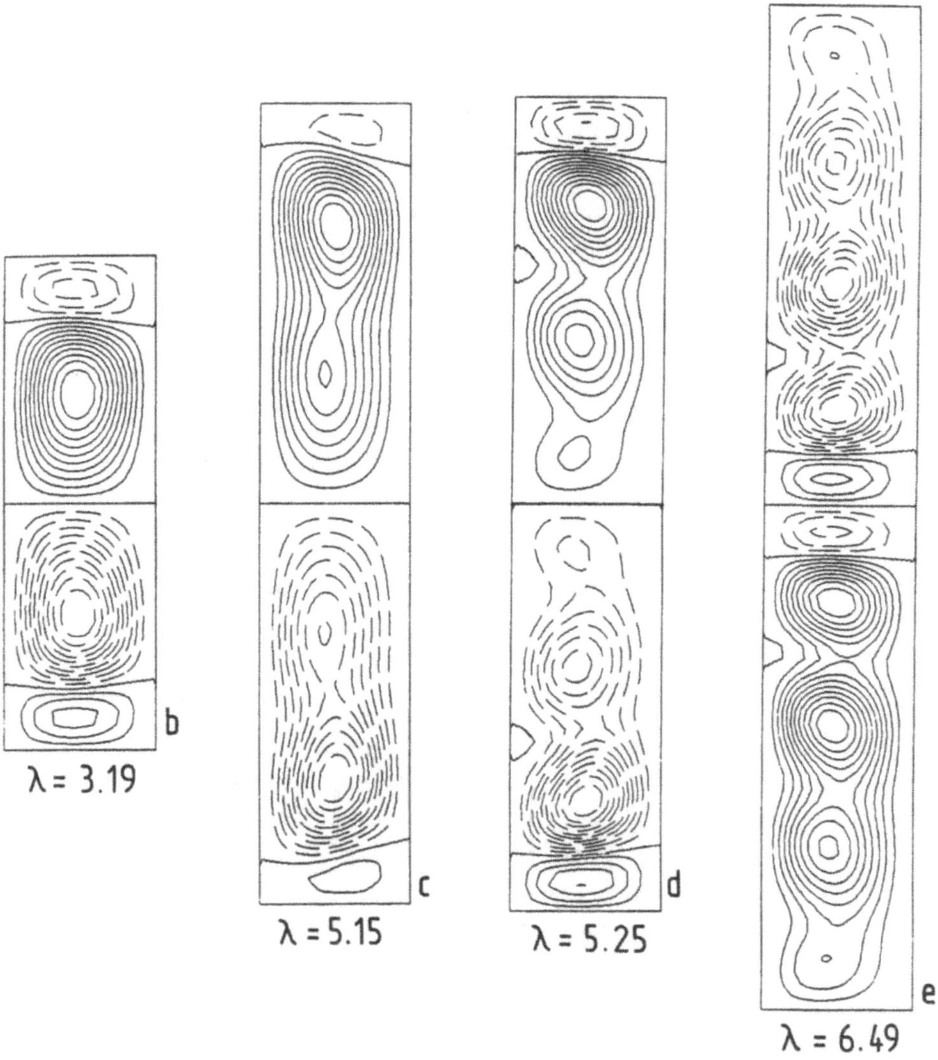

Figure 3.32: Level lines of the stream function ψ for solutions on the different isolas shown in Fig. 3.26. Computations with $N = 14$, $\varepsilon < 10^{-1}$.
b) Outward transition solution on the branch shown in Fig. 3.26b. It lies between the solutions shown in Figs. 3.28a and b. *c*) First fold point beyond the ×4-bifurcation $\lambda \to 4\lambda$ on the isola shown in Fig. 3.26c. Tests with $N = 8$ and $N = 12$ gave essentially the same results. *d*) Solution between ×6-bifurcation and first fold point on the isola shown in Fig. 3.26d. Computations with $N = 8$ and $N = 12$ yielded essentially the same results. *e*) Solution between the ×8-bifurcation and the perturbed bifurcation on the isola of Fig. 3.26e. [SWM, Fig.10].

flow in a driven cavity, as shown in [Nea87, Fig. 3], with the flows of Figs. 3.28 g, h.

Figure 3.26 still does not give a full summary of the flows existing for $\mathrm{Re} = 2\,\mathrm{Re}_{cr}$ in the λ-interval investigated. There are many more. For instance, we found a twin pair of isolas bifurcating from the mixed-mode part of the $\times 3$-branch. They are shown in Fig. 3.33. Their kinetic energies are included in Figs. 3.26 f_3, f_4. The pair of $(\lambda \to 3\lambda)$ bifurcations at $\lambda = 5.38$ might be tertiary bifurcations generated by an interaction of secondary bifurcations on the $\times 6$-branch. Known secondary bifurcations on the $\times 6$-branch are:

i) $\lambda \to 2\lambda$, generated by the $(6,12) \equiv (2,4) \times 3$ double point (second periodic repetition of the $(2,4)$ interaction). For $\mathrm{Re} = 2\,\mathrm{Re}_{cr}$ it occurs at $\lambda = 1.787 \times 3 \approx 5.36$ (Fig. 3.27e). It is included in Fig. 3.33 as the $\lambda \to \lambda/2$ bifurcation on the $\times 3$-branch.

ii) $\lambda \to 3\lambda$, generated by the $(4,12) \equiv (2,6) \times 2$ double point. For $\mathrm{Re} = 2\,\mathrm{Re}_{cr}$ it occurs at $\lambda = 2.6459 \times 2 \approx 5.29$ (Fig. 3.28e).

iii) $\lambda \to 6\lambda$, generated by the $(2,12)$ double point. For $\mathrm{Re} = 2\,\mathrm{Re}_{cr}$ it occurs at $\lambda = 4.98$ [WS87, Abb. 86].

In Fig. 3.34 we show more flows, to give an idea of the variety to be expected. These flows were obtained by pseudo-arclength continuation in decreasing Re and fixed $\lambda = 6.8$ from solutions on the three branches for $\mathrm{Re} = 800$ shown in Fig. 3.11b, c.

A more systematic way to obtain these flows and others would be: Follow all primary branches bifurcating from Couette flow at the curves shown in Fig. 3.25 by pseudo-arclength continuation, and study their nonlinear interactions, and the nonlinear interactions of the branches generated in these interactions, etc.. Figure 3.29 shows the beginning of such an analysis for $\lambda = 3.4,\ \mathrm{Re} \leq 2\mathrm{Re}_{cr}$. Because of the asymmetries in the Taylor problem (and thus *perturbed* bifurcations, *transcritical* bifurcations) such an analysis soon becomes complex and cumbersome. Figure 3.18 suggests that it is more efficient to study these interactions in the Rayleigh-Bénard problem first. No bifurcations are perturbed, and they are all symmetry-breaking and thus simple pitchforks. When this work is done, symmetry considerations along the lines of [TC, Wer, Bus87] will show what happens to these bifurcations in the Taylor problem, in the Benjamin problem, or for flows between spheres. Investigations in this direction are discussed in the next chapter.

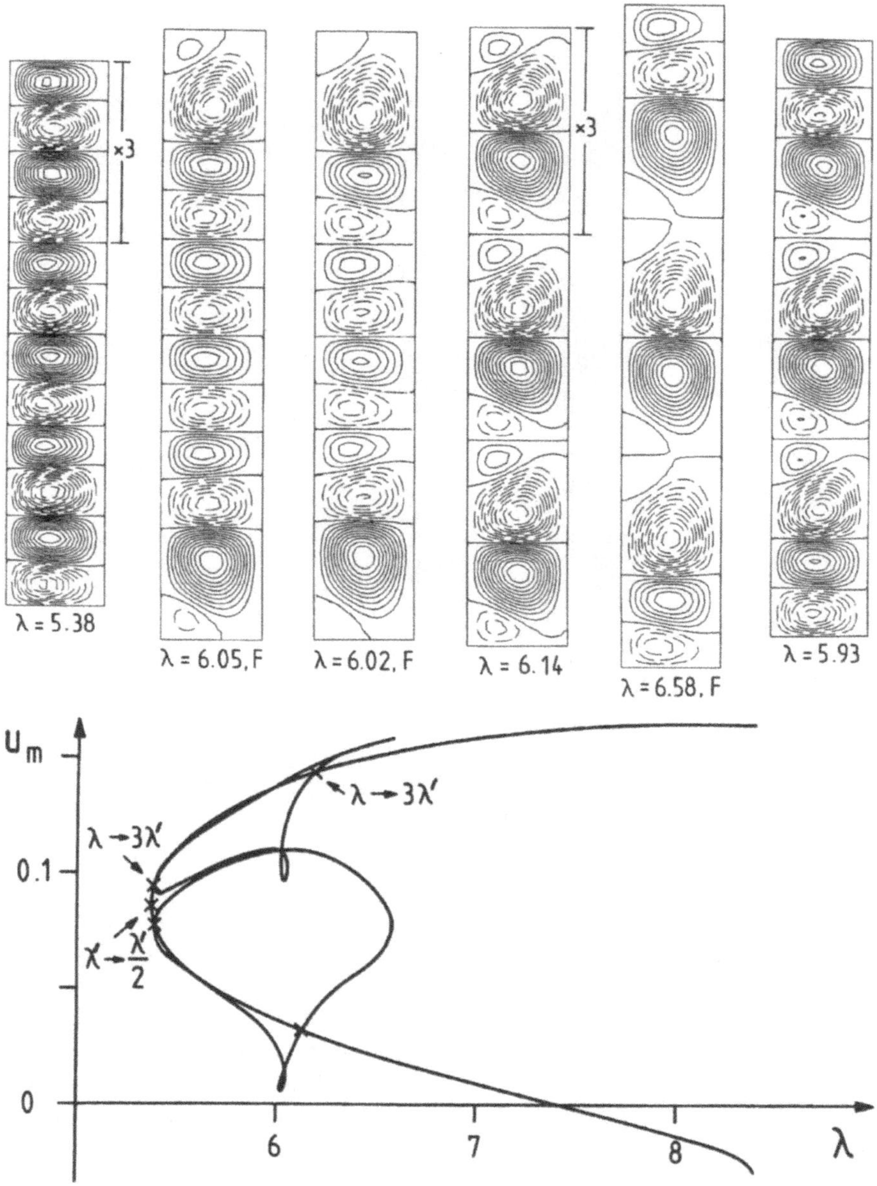

Figure 3.33: Twin pair of isolas bifurcating from the mixed-mode part of the ×3-branch near the periodic repetition of the $\lambda \to \lambda/2$ bifurcation shown in Fig. 3.27e. Level lines of ψ for flows on the upper isola, proceeding counter-clockwise. Computations with $N = 14$, $\varepsilon < 10^{-1}$. $u_6(r_m, z_m; \lambda)$ is the maximum Fourier coefficient [SWM, Fig.11].

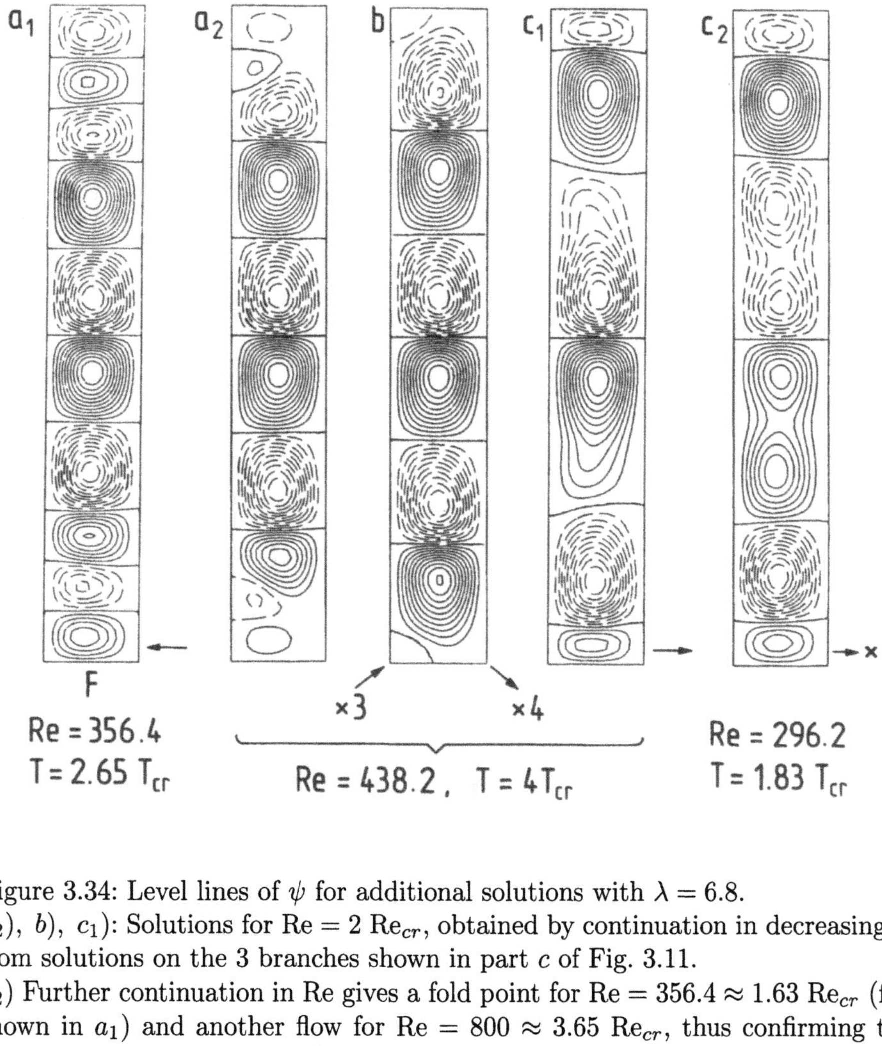

Figure 3.34: Level lines of ψ for additional solutions with $\lambda = 6.8$.
a_2), b), c_1): Solutions for Re $= 2$ Re_{cr}, obtained by continuation in decreasing Re from solutions on the 3 branches shown in part c of Fig. 3.11.
a_2) Further continuation in Re gives a fold point for Re $= 356.4 \approx 1.63$ Re_{cr} (flow shown in a_1) and another flow for Re $= 800 \approx 3.65$ Re_{cr}, thus confirming that Fig. 3.11c is not complete. b) For Re $= 654 \approx 2.98$ Re_{cr} this branch meets the 'pure-mode' part of the $\times 3$-branch, and for Re $= 226 \approx 1.03$ Re_{cr} it ends on the 'pure-mode' part of the $\times 4$-branch. Computations with $\lambda = 6.5$ yielded essentially the same results. c_1) This solution was obtained from a flow between the flows numbered '24' and '25' in Fig. 3.12. When decreasing Re further, we arrived at the flow c_2, and then, for Re $= 266 \approx 1.21$ Re_{cr}, in a bifurcation to the 'pure-mode' part of the $\times 2$-branch. Computations with $\lambda = 6.5$ had a bifurcation to the 'pure-mode' part of the $\times 5$-branch for Re $= 255 \approx 1.16$ Re_{cr}. This can happen because $6.5 < \lambda_{4,10} < 6.8$, see Fig. 3.25. Computed with $N = 12$, $M = 31$ [SWM, Fig.12].

3.6 Stability of the stationary vortices

Until now, our attention was focused on mathematical properties of the flows. The important question of stability was discussed only briefly. Here we shall report on stability investigations by other authors. It will become obvious that the bifurcation structures investigated in the previous sections strongly influence the stability of stationary vortices in wide gaps.

Depending on η, two different types of instabilities in stationary Taylor vortex flows for $\mu = 0$ are known: The emergence of wavy vortices and the Eckhaus instability/short-wavelength instability. We shall discuss them in this section.

3.6.1 Wavy vortices

The transition from (purely toroidal) Couette flow to axisymmetric stationary vortices was first described by Taylor [Tay]. The linearized stability theory of this transition is well understood and was discussed in Chap. 1. The next transition observed was the one when the Taylor vortices become wavy, i.e. unstable to non-axisymmetric disturbances. The torque versus Reynolds number curve changes its inclination when this instability occurs (see Fig. 1.3).

The transition to wavy vortices was investigated theoretically by Davey, DiPrima and Stuart [DDS] in 1968. Their technique was based on expanding the solution in powers of the parameter $\mathrm{Re} - \mathrm{Re}_{cr}$, which was assumed small[9]. This procedure was carried further by Eagles [Eg71], who also used it to find the torque on the cylinders [Eg74] and thus confirmed that the change of inclination is caused by this instability. These workers considered finite amplitude perturbations, both axisymmetric and non-axisymmetric, to the state of Couette flow. This method works well just above the neutral curve, but is not adequate when the parameter $\mathrm{Re} - \mathrm{Re}_{cr}$ is not small (see for instance [KS69, Figs. 1-3] or [DPE, Fig. 2]).

Numerical methods thus seem necessary, because these restrictions do not apply to them. The first numerical investigations of wavy vortices were probably those by Ken Meyer [Mey69]. He crudely approximated wavy vortex flows with a square grid of 9 to 14 grid points in the (r, z)-plane and one Fourier mode in the toroidal direction. He checked numerically the analytical linear [Rob65] and nonlinear [DDS] stability analysis by studying

[9] Kirchgässner (in his Habilitationsschrift 1966) established rigorously that stationary Taylor vortex flows have an expansion in powers of $(\mathrm{Re} - \mathrm{Re}_{cr}) \sim (T - T_{cr})^{1/2}$.

growth rates for small three-dimensional disturbances to Couette flow and to stationary Taylor vortex flows , and he contributed to the question if the torque reduction observed in the experimental data displayed in Fig. 1.3 is caused by the flow becoming wavy at these parameter values. He showed that the torque of stationary vortices decreases for decreasing wave number α, but that $\alpha = 1.8$ ($\lambda \approx 3.5$) would be necessary to match the measured torque values, and such large wavelengths were not observed. This was then continued by Eagles [Eg74] who confirmed that wavy vortices have smaller torque than stationary vortices of the same wavelength.

A more complete picture of the wavy-vortex instability was given by Jones [Jo81], who investigated it in detail and clarified its dependence on η. He did a linear stability analysis in a neighborhood of stationary solutions as is appropriate in the nonlinear regime. The two-dimensional stationary 2-vortex solutions were approximated by a Galerkin method, with Fourier expansion in the axial direction and Čebyšev polynomials in the radial direction. Their stability was investigated by considering small non-axisymmetric perturbations of these 2-vortex solutions and then finding the eigenvalues of a matrix related to the differential equations. This method had been used before for the Rayleigh-Bénard problem, by Clever, Busse [CB74] and Jones, Moore [JM79]. It is valid near the stability boundary of the onset of waviness for all η values and has the advantage that it is not necessary to solve three-dimensional partial differential equations. It permits some exploration of the high Re-number regime for those η-values for which this seems necessary: the computations were accurate for better than 1% up to Re $\approx$ 5 Re_{cr}. The frequencies found by Jones for the unstable modes were in good agreement with the experiments by Coles ($\eta = 0.874$) and Donnelly ($\eta = 0.8756$).

Jones' main results are visualized in Figure 3.35. This figure shows the marginal curves for onset of wavy vortices with toroidal wave numbers $m = 1, 2, 3$ for $\mu = 0$, $\alpha = 3.13$ (i.e. $\lambda \approx 2.007$). Other toroidal wave numbers m were investigated as well, and were found to be less important. Figure 3.35 shows that the stability behavior of axisymmetric vortices changes substantially for $\eta_{wcr} \approx 0.75$: for $\eta > \eta_{wcr}$ wavy vortices occur for Re-numbers close to Re_{cr}, for $\eta < \eta_{wcr}$ flows stay stationary and axisymmetric up to much higher Re-numbers. It should be noted that the onset of waviness is strongly λ-dependent: for $\eta = 0.75$, the transition to wavy vortex flows occurs at Re $= 2$ Re_{cr} for $\lambda = 2.62$, and at Re $= 6$ Re_{cr} for $\lambda = 1.8$. As explained earlier, we chose the value $\eta = 0.727$ for our

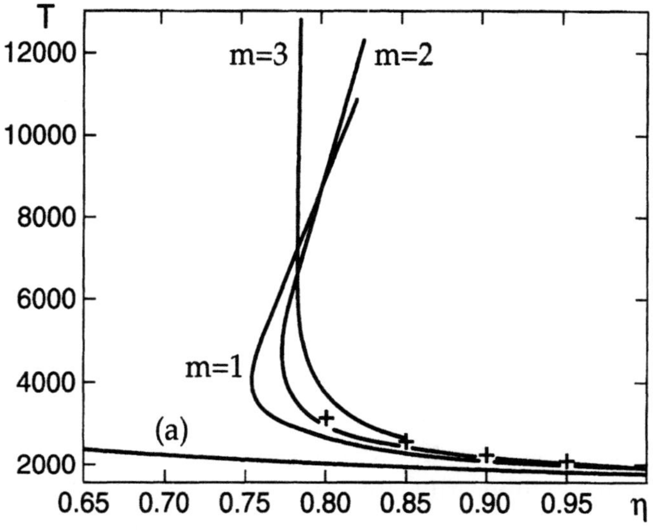

Figure 3.35: Neutral curves in the (η, Taylor number)-plane for $\mu = 0$ and axial wave number $\alpha = 3.13$, i.e. $\lambda \approx 2$.
Curve (a) is the neutral curve for the onset of axisymmetric 2-vortex Taylor vortex flows. The curves $m = 1$, $m = 2$ and $m = 3$ are curves of neutral stability for the non-axisymmetric modes with azimuthal wave number m. In the region above and to the right of the envelope of these curves the vortices are wavy [Jo81, Fig. 3].

investigations of stationary flows because of Jones' analysis, and because Burkhalter, Koschmieder observed stationary flows up to Re $>$ 9 Re_{cr} for various wavelengths [BuKo].

In Chap. 1, we defined 'narrow gaps' as gap widths for which the narrow gap approximation (1.15) is valid. Jones' analysis suggests that those gaps should be considered narrow, for which stationary axisymmetric vortices lose stability to wavy vortices for $\text{Re}/\text{Re}_{cr} \approx 1$, i.e. all gaps with $\eta > \eta_{wcr} \approx 0.75$. This led us to call $\eta = 0.727$ a 'narrow wide gap' in an earlier section.

Further numerical investigations of wavy vortices were performed in the meantime. We shall not discuss them here, but mention them briefly: Computations by Marcus [Ma84] who compared his results with those of Meyer-Spasche, Keller in the testing phase of his code, and computations by Schröder, Keller [ScK]. When wavy vortices become unstable, the flow becomes quasiperiodic under certain conditions. Such quasiperiodic flows were investigated theoretically and numerically by Coughlin, Marcus [CoM92].

3.6.2 Eckhaus and short-wavelength instabilities

When the marginal curve for stability of Couette flow was found, the common belief first was that Taylor vortex flows exist everywhere in the region above the envelope of this curve and its periodic repetitions, and are stable there, at least in the case $\mu \geq 0$. As we have seen in earlier sections, there are cases in which Taylor vortex flows exist outside this region because of subcritical bifurcations, and also portions of this region where they don't exist because of the basic $(n, 2n)$-fold. As we shall see now, there are portions of this region where they exist, but are not stable.

For Reynolds numbers just above Re_{cr}, stationary flows lose stability due to the **Eckhaus instability**. This is an axisymmetric long-wavelength modulation of the pattern: If a wavelength becomes 'too long', the stationary flow becomes temporarily time-dependent, adjusts the wavelength to a shorter value inside the Eckhaus limit, and becomes stationary again. This adjustment is done by creation of one additional pair of vortices which grow until all vortices have a shorter wavelength. On the other hand, when the Eckhaus limit is reached from the short wavelength side, a pair of vortices disappears, and a new stationary state with longer vortices is reached [Can]. The Eckhaus instability is thus a long-wavelength bulk instability: it is a global instability; all vortices change their size simultaneously.

The stability of stationary flows to axisymmetric perturbations was first investigated by Kogelman, DiPrima [KD]. They generalized a perturbation approach introduced by Eckhaus [Eck65, DES71][10]. The curve of marginal stability for Taylor vortex flows obtained by their analysis is included in several of the following figures, e.g. Fig. 3.37. Their approach used expansions up to third order and is thus only valid in a neighborhood of the marginal curve of Couette flow. Moreover they excluded several effects which we found typical for wide gap problems: for instance they assumed that the marginal curve divides the (λ, Re)-plane into a region where Taylor vortex flows exist (above) and don't exist (below). As can be seen from Fig. 3.40, their approximation ceases to be valid as soon as secondary bifurcations and the basic $(n, 2n)$-fold play a role.

Numerically, the stability was investigated by Riecke and Paap [RP86, RP90] for $\mu = 0$ and $\eta = 0.892$, 0.75 and 0.5. Their work probably gives the first detailed, quantitative agreement of theoretical and experimental Eckhaus limits (Fig. 3.37). They used methods influenced by and similar

[10] Eckhaus (1965) investigated the later-named 'Eckhaus instability' in connection with the nonlinear Orr-Sommerfeld problem for Poiseuille flow.

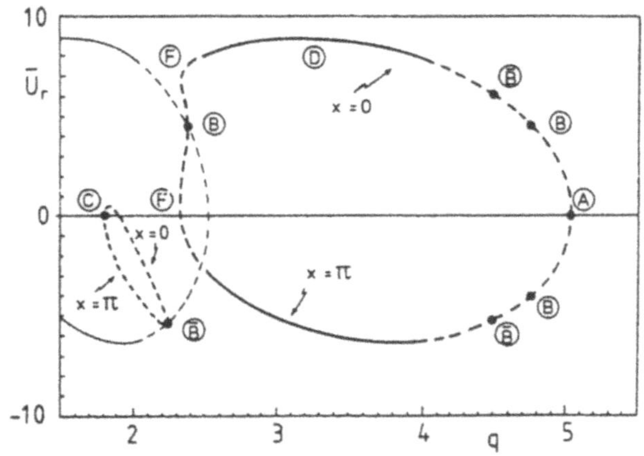

Figure 3.36: Stability of Taylor vortex flows in the solution diagram velocity u_m vs. wave number $q = 2\pi/\lambda$ for $\eta = 0.75$, $\mathrm{Re}^2 = 1.195^2\ \mathrm{Re}_{cr}^2 \approx 1.43\ \mathrm{Re}_{cr}^2$; Points $A = (1)$, $B = (5)$, $\bar{B} = (4)$, $C = (3)$, $F = (6)$ in Fig. 3.4; twin points and periodic repetitions have the same names. —: stable; - - -: unstable [RP86, Fig. 4a].

to the methods Jones used in his investigations of the wavy-vortex instability: computation of stationary vortices using a Galerkin method (Fourier expansion in the axial direction and Čebyšev polynomials in the radial direction) and investigation of their stability to *axisymmetric* perturbations by finding the growth-rate σ of the instability as an eigenvalue of a matrix related to the differential equations. As described in the previous section, Jones' investigations were influenced by Clever, Busse [CB74]. These authors investigated many different instabilities in the Rayleigh-Bénard problem, including the Eckhaus instability. In convection layers, the Eckhaus instability is important in fluids with low Prandtl number (4.9), $P \leq 1.1$: more important in mercury ($P = 0.025$) than in air ($P = 0.71$); not occurring in water ($P = 7$) [CB74, Figs. 5, 6, 7].

The results of the analysis of Riecke, Paap are illustrated by Figs. 3.36, 3.37, 3.39 and 3.40. Figure 3.36 shows the limits of stability for Taylor vortex flows in the solution diagram of Fig. 3.4 for slightly different values of η and Re. At first glance it is surprising that the loss of stability is not related to the fold point or a bifurcation point. This is explained by the fact that the Eckhaus instability is a global, long wavelength instability: a local analysis using periodic boundary conditions and a simplified set of equations cannot detect it by such simple means as a sign change of the

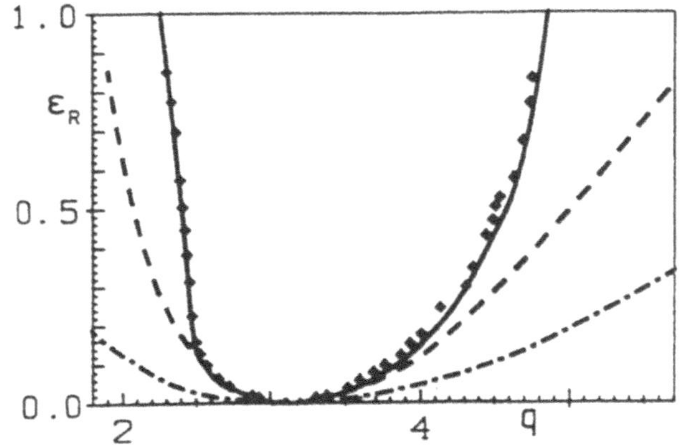

Figure 3.37: Marginal curve for stationary Taylor vortex flows in the (q, ε_R)-plane, $\varepsilon_R := (\mathrm{Re} - \mathrm{Re}_{cr})/\mathrm{Re}_{cr}$ and $q = 2\pi/\lambda$, for $\eta = 0.75$, $\mu = 0$.
—: Computations by Riecke and Paap [RP86]; ◇◇◇: experiments by Domingues-Lerma et al ([ACDH] and unpublished); – – –: 3rd order perturbation calculations by Kogelman and DiPrima [KD]; – · –: marginal curve of Couette flow [RP86, Fig. 3].

determinant: the computational domain itself must be allowed to change when this instability sets in. Tests supplementing the stability analysis confirmed that the loss of stability is also not related to the change of structure on the solution branch: for several parameter values mixed-mode solutions containing large and small vortices were found to be stable [Rieck]: stable and unstable flows do not 'look' different on the two sides of the Eckhaus limit (also, flows that are stable or unstable to wavy-vortex perturbations do not 'look' different).

Nevertheless it would be wrong to say that the bifurcation structure displayed in Figs. 3.36 and 3.4 is unimportant for the Eckhaus stability limit. If we compare the stability limits as given from the third order amplitude expansion and from the numerical stability analysis (Fig. 3.37), we see that the amplitude expansion gives good results close to threshold ($\mathrm{Re} < 1.1\ \mathrm{Re}_{cr}$), as is expected. For larger $\mathrm{Re} - \mathrm{Re}_{cr}$ the expansion does poorly, especially on the low-q (high λ) side. The authors report that this is not due to a change in the type of instability, but to the bifurcation structure of the solution set: in computations for fixed Re, the longitudinal phase diffusion constant $D_{\|}$ diverges at the fold point due to terms involving derivatives with respect to $q = 2\pi/\lambda$, and the growth rate σ was

Figure 3.38: Same as Fig. 1.8, but upside down. This confirms that flows at the short wavelength Eckhaus limit are more sensitive to the influence of gravity than flows with longer wavelengths [BuKo, Fig.8].

found to increase quite rapidly when the fold point was approached. This rapid growth probably allows such a good agreement of computations and measurements as can be seen in Fig. 3.37.

In the beginning, the agreement between computations and experiment was not nearly as good on the high q side as on the low q side [AC]. It was found that this was not only due to the fact that the diffusion constant $D_{\|}$ does not diverge near the Eckhaus limit on the high-q side. Agreement could be improved by using horizontal instead of vertical cylinders. The experimenters found that gravity played a role on the flow of a two-component fluid (oil with dissolved aluminum particles): '... the vortices near the bottom became compressed and those near the top were expanded'. If we look at Fig. 1.8 again, we can see this a little bit for the flow with shortest λ_f. Putting the same figure upside down as in Fig. 3.38 magnifies the effect: the wavelength varies continuously in the flow on the r.h.s., while the wavelengths of the other two flows seem to be constant along the whole lengths of the cylinders. Flows at the short wavelength Eckhaus limit are more strongly affected by the influence of gravity than flows with longer wavelengths. It also shows that our brain is quite capable of suppressing the effect of gravity in our way of viewing the world: the picture has to

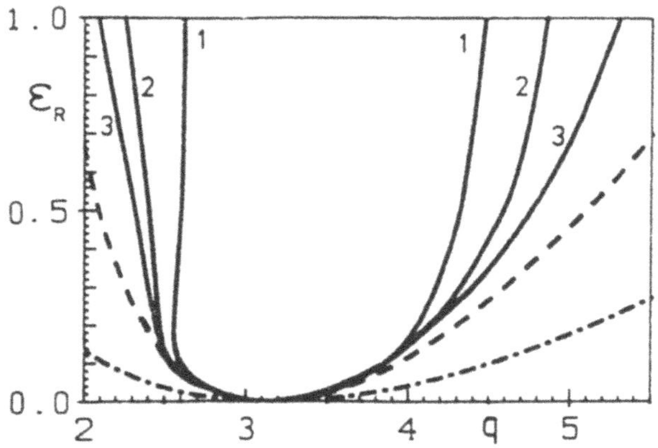

Figure 3.39: Comparison of Eckhaus limits for several ηs ($1 : \eta = 0.5$, $2 : \eta = 0.75$ and $3 : \eta = 0.892$). Also shown are the Eckhaus limit to third order [KD] and the marginal curve of Couette flow (both are nearly independent of η) [RP86, Fig. 7].

be put upside down to make us see this effect. Gravity is neglected in the mathematical model, and only one fluid is considered.

Figure 3.39 shows the dependence of the Eckhaus limit on η. It uses the fact that the marginal curve of Couette flow is nearly independent of η for Reynolds numbers normalized by the critical number $\mathrm{Re}_{cr}(\eta)$, as was explained at the beginning of the section 'interactions of secondary branches'. Note that for $\eta = 0.892$ Taylor vortex flow is already unstable to non-axisymmetric perturbations for $\mathrm{Re}/\mathrm{Re}_{cr} > 1.12$, i.e. for $\varepsilon_R > 0.12$.

How does Fig. 3.37 change when the Reynolds number is increased? From the computations reported earlier (Fig. 3.21) we know that the fold becomes more and more pronounced. Riecke, Paap [RP90] report that the two Eckhaus limits approach the points $F = (6)$ and $\bar{B} = (4)$ for increasing Re, and that a different instability replaces the Eckhaus instability when the point $\bar{B}$ has been reached. For $\eta = 0.5$, this happens at $\mathrm{Re} = 2.54\ \mathrm{Re}_{cr}$. The new instability affects *every second* vortex pair and diminishes it. They call this instability the **short-wavelength instability**. It seems to be similar to the Eckhaus instability in that the flow becomes time dependent only temporarily, and then it settles at a stationary flow with different wavelength, with double the old wavelength this time: it seems to be a 4-2 transition. The location of the marginal curve for the short-wavelength instability can be seen from Fig. 3.40. Also, this figure shows the dependency of the Eckhaus limits on the bifurcation structure of the solution set.

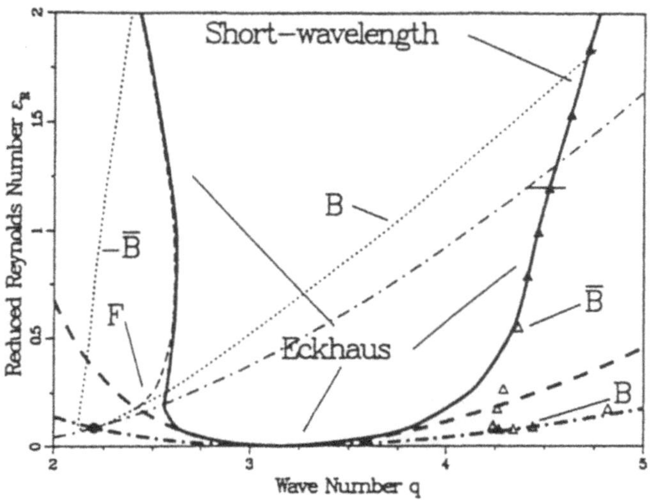

Figure 3.40: Eckhaus and short-wavelength limits for $\eta = 0.5$, $\mu = 0$.
——: full stability analysis; - - -: 3rd order amplitude expansion [KD]; —·—: marginal curves of Couette flow for 2-vortex and 4-vortex flows; •: the $(2, 4)$ double point; ···: points $B = (5)$, $\bar{B} = (4)$ and $F = (6)$ as functions of $q = 2\pi/\lambda$ and ε_R. Δ Δ Δ: periodic repetitions of (4) and (5) for $2q = 4\pi/\lambda$ [RP90, Fig. 5].

Chapter 4

Secondary Bifurcations on Convection Rolls

As we pointed out in Chapter 1, the equations of the Boussinesq approximation describing convection rolls can be seen as a limiting case of the equations of the Taylor problem. In this chapter we give the algorithm for performing this limit.

Then we investigate the nonlinear interaction of 2-roll solutions with 4-roll and with 6-roll solutions. Fourier expansion, the Galerkin method and perturbation arguments are used to reduce the PDEs of the Boussinesq approximation to several different nonlinear algebraic systems. For these systems we calculate and compute the loci of secondary bifurcations. This verifies several conjectures about the global structure of secondary bifurcations formulated in Chapter 3. In addition, it gives a more complete picture of pattern formation and allows new conjectures.

4.1 Introduction

In the numerical investigations of the Taylor problem reported in the previous chapter, we found strong evidence that a curve of secondary bifurcations on the 4-vortex solution branch connects two different double points, i.e. two different intersection points of primary bifurcation curves in the (λ, Re)-plane for fixed η and μ. It seemed desirable but too complicated to confirm this result by analytic investigation of the Taylor problem, and hopeless to get further results by generalization. The many perturbed and transcritical bifurcations in the Taylor problem make it difficult to understand bifurcation patterns.

As we have seen near Figs. 3.17 and 3.18, the fold points on the basic $(n, 2n)$-fold correspond to bifurcation points on convection roll branches. And the subcritical fold does not exist on convection rolls since the bifurcations are supercritical. This is so because the Rayleigh-Bénard problem is more perfect than the Taylor problem: It has a symmetry with respect to the mid-plane between the horizontal plates, see eq. (4.68). This symmetry is broken in the Taylor problem. Also, it is possible in the Rayleigh-Bénard problem to use stress-free boundary conditions in the vertical direction, and it is known that stationary rolls are essentially the same for rigid and stress-free boundary conditions. The stress-free boundary conditions in the vertical direction together with periodic boundary conditions in the horizontal direction allow Fourier decomposition in both space dimensions. An analytic treatment of the Rayleigh-Bénard problem is thus less complicated and promises more insight than the corresponding investigations for the Taylor problem. The relation between the equations of the Taylor problem and those of the Rayleigh-Bénard problem in the Boussinesq approximation can thus be employed to get more insight on the solution manifold in the Taylor problem. This will be pursued in this chapter.

First we briefly discuss convection movements occurring in nature. In many cases these can be described by the equations of the Boussinesq approximation derived from the same basic equations of continuum mechanics which can be used to derive the Navier-Stokes equations. Rigid and stress-free boundary conditions are discussed. We then show how (in certain parameter regimes) the equations of the Boussinesq approximation with rigid boundary conditions on top and bottom of the fluid layer can be seen as a limiting case of the equations modeling the Taylor problem. We give an algorithm-like transformation for the full 2-dimensional (2D) equations in primitive variables. This transformation could be used to write a Taylor-Rayleigh-Bénard code for numerical investigation of stationary 2-dimensional solutions to both problems.

Then we introduce the stream function ψ and stress-free boundary conditions on top and bottom of the fluid layer. This system of equations is Fourier decomposed with respect to both space dimensions. Then it is easy to calculate the neutral curves $R_{cr}^{p1}(\alpha)$, $p = 1, 2, 3, \ldots$ of the primary solution in the (α, Ray)-plane explicitly, and also the pure-mode secondary solutions bifurcating at these curves. Here α is the wave number in the horizontal direction of the fluid layer.

To get analytic results on secondary bifurcations, it seems necessary to use some simplified version of the Fourier decomposed system.

There are several local investigations of secondary bifurcations on convection rolls [Bus87, BO, KG83, Seg62]. Essentially, all these authors use a Fourier decomposition in both spatial coordinates and an amplitude expansion to 3rd order in $(\alpha - \alpha_0)^{1/2}$ and $(\text{Ray} - \text{Ray}_0)^{1/2}$, where α_0 and Ray_0 are those values of the wave number and of the Rayleigh number, respectively, for which the neutral curves of two different numbers of rolls intersect. The analysis is restricted to a neighborhood of the double point (α_0, Ray_0), i.e. to a neighborhood of one of the two interacting double points of simultaneous interest to us.

To investigate the nonlinear interaction of perturbations of the wavelengths λ_1 and λ_2 to third order, the modes corresponding to λ_1, λ_2, $\lambda_1+\lambda_2$ and $\lambda_1 - \lambda_2$ have to be taken into account. The resulting system involves 9 equations [Seg62, KG83]: the investigation of the interaction of the $(k,1)$-mode with the $(m,1)$-mode shows that the modes $(k,1)$, $(m,1)$, $(k+m,2)$, $(m-k,2)$, $(0,2)$ should be taken into account. In the special case of the interaction of 2-roll solutions with 4-roll solutions, this leads to $\lambda_1(\alpha) = 2\pi/\alpha, \quad \lambda_2(\alpha) = 4\pi/\alpha$ in the horizontal x-direction and $\lambda_1 = \lambda_2 = \pi$ in the vertical z-direction. It corresponds to the Fourier modes 11, 21, 02, 12 and 32.

The two organizing centers for the interaction and pattern forming (the double points $(\text{Re}_{24}, \lambda_{24})$ and $(\text{Re}_{d4}, \lambda_{d4})$ defined near Fig. 3.24) seem to stay distinct for all parameter values in the Taylor problem. In the Rayleigh-Bénard problem, it can be seen from the simple formulas (4.108) for the intersecting curves that the two corresponding double points cannot merge for any parameter value. Local analysis by perturbation method in the neighborhood of a point thus is not adequate for investigating their interaction.

For *global* investigation of mode interactions we consider several model problems and methods:

a) The first model problem consists of 7 equations. It was introduced by us [LMP] for investigation of the 2-roll, 4-roll interaction. The loci of secondary bifurcations are found by calculation of vanishing determinants in dependence on the three parameters in the problem.

To get analytic results which are valid simultaneously in a neighborhood of both double points corresponding to $(\mathrm{Re}_{24}, \lambda_{24})$ and $(\mathrm{Re}_{d4}, \lambda_{d4})$, we [LMP] used the following approximation to the Fourier decomposed system: the number of Fourier modes is restricted from the very beginning to the first two modes in both directions (Galerkin method). This gives 10 equations. Perturbation arguments are then used for deciding which other terms and equations can be neglected without loss of important information in an (α, Ray)-neighborhood of the neutral curves. This reduces the system to 7 equations. These can be solved explicitly for some of the pure-mode solutions of the full system which were mentioned earlier. The secondary bifurcations on these solutions are calculated by evaluating 7×7 determinants.

The Galerkin approach was justified with the following reasonings:

1) As we have seen in Chapter 2, the extensive studies of the numerical accuracy of the code Tayperio have shown that the contribution of 3rd and higher axial modes is negligible in the study of the interaction of the 1st and 2nd mode. Unpublished computations even gave the impression that it is advantageous to exclude higher modes in the study of the *basic* pattern. This is supported by Fig. 2.11.

2) In their paper 'Multiple eigenvalues lead to secondary bifurcations' Bauer, Keller, Reiss [BKR] discuss an example in which the two interacting pure-mode solutions are orthogonal to each other in the underlying function space, and the secondary mixed-mode solutions are represented by linear combinations of these pure-mode solutions. Since we are dealing with a Fourier decomposed PDE here, the mixed mode solutions are more complex: the contributions of the 12 and 02 Fourier modes turn out to be essential for the results.

As explained earlier, a local perturbation analysis to 3rd order shows that the 32-modes are essential as well. It has been suggested that neglecting them could produce a wrong curvature of the center manifold in one or both of the double points involved and thus to give a wrong inclination of the curve of secondary bifurcations in these double points [K91]. To clear this point, we also investigate model c). It turns out that the neglected modes do change the results, but these changes are negligible on a global scale.

b) The second model problem consists also of 7 equations. It was introduced by us [Mat90] for investigation of the 2-roll, 6-roll interaction. Model and method are analogous to case a) and help to generalize to the 2-roll, n-roll case. The determinants to be evaluated for finding the loci of secondary bifurcations turn out to be very similarly structured as in the 2-roll, 4-roll case. The assumption that there are again two interacting double points and that everything will be essentially analogous is confirmed.

The results on the 2-roll, 6-roll interaction remained unpublished until now because it was felt that the error introduced by the Galerkin method should be investigated for the Rayleigh-Bénard problem as well. Instead of the original reasoning in [Mat90] which modes to neglect for model b), a reasoning on the background of model c) is used here. It renders the same result, is simpler and gives more insight.

A comparing view on the secondary bifurcations generated by the 2-roll, 4-roll and the 2-roll, 6-roll interactions confirms the possibility of nonlinear interactions between the (2,4)- and the (2,6)-tertiary branches bifurcating from the 2-vortex solutions, and locates an organizing intersection point in the (α, Ray)-plane in case of the Rayleigh-Bénard problem. Such interactions were detected numerically (see Figs. 3.28 to 3.33 and their discussion).

c) The third model problem consists of 9 equations for the 2-roll, 4-roll interaction. It is obtained from system a) by including the 32-modes. The loci of secondary bifurcations are found by computation of eigenvalues with standard software and iteration for vanishing eigenvalues. The same computations are applied to model problem a). Comparison allows to estimate the error introduced by the Galerkin method in model a) and shows that it is small on a global scale. In addition, these computations confirm that the results of the tedious calculations of determinants are correct in the version given here.

A comparison of the 7 equations used in the global model a) for the 2-roll, 4-roll interaction with the 9 equations used by Knobloch, Guckenheimer [KG83, eqs. (15), (16)] in their local perturbation analysis to third order shows:

1) All modes and equations are the same except that two are missing in model a). These two are included in model c).

The Fourier modes neglected in model a) of the 2-roll, 4-roll interaction are the 32-modes, which seemed to be unimportant by numerical experience (see above). A detailed comparison shows that the 32-modes always change the results, but these changes are negligible on a global scale, see section 'The perturbation approach' and Figs. 4.6, 4.7 and 4.8.

2) Model c) describes a full neighborhood of several intersecting curves of primary bifurcations in their dependence on 3 parameters (α, Ray, Pr), not only a neighborhood of one intersection point.
This is due to the fact that we use the perturbation arguments only for neglecting modes and equations, but we do preserve the dependence of the original equations on the parameters α and Ray instead of expanding in a neighborhood of a double point.[1]

Then we discuss how the results obtained for model problems a) to c) can be generalized. And we report on computations by Bolstad, who followed several of these singularities with a fold following and bifurcation point following algorithm in the (Re, λ)-plane.

This chapter shows that the conjectures formulated in the previous chapter are confirmed both in the narrow gap limiting case of the Rayleigh-Bénard problem and for wide gaps ($\eta = 0.727$ and $\eta = 0.615$). Thus it seems reasonable to assume that the global bifurcation structures found essentially hold for all parameter values and probably are a general feature of the Navier-Stokes equations.

4.2 The Rayleigh-Bénard problem

4.2.1 Convection in fluids

Consider a horizontal layer of a fluid, hotter on the lower side than at the upper. A very small temperature difference is reduced by heat conduction only. Above a critical temperature difference, however, viscosity and thermal conductivity cannot stabilize anymore this state with lighter fluid below heavier fluid. It becomes unstable due to buoyancy: convection sets in.

This convective instability is often encountered in nature.
If convection occurs in the atmosphere of the earth, it leads to the appearance of cumulus clouds.
Since the magma in the liquid kernel of the earth does have electro-magnetic properties, geophysicists assume that convection of this magma generates the earth's magnetic field. This is dealt with in the dynamo theory which tries to explain the origin of the magnetic fields of earth and planets through the dynamo effect [Bus83, LM82].

[1]The Lorenz equations were obtained by a similar procedure for the time dependent system, using only the 11 and 02 modes [GH, p. 92].

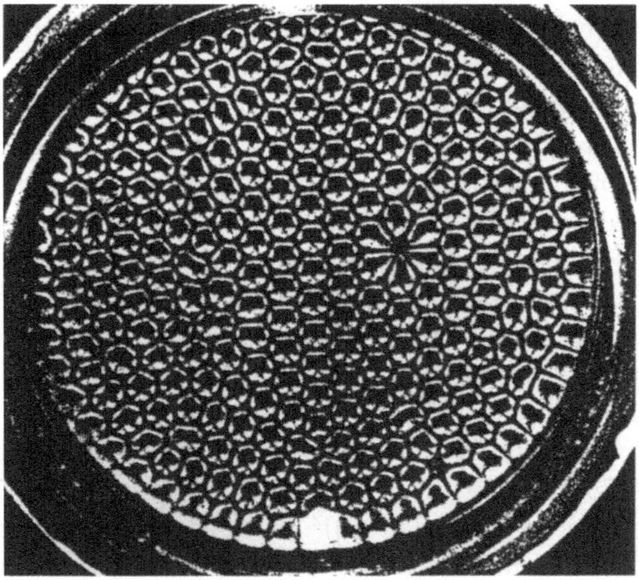

Figure 4.1: Hexagonal convection cells in an experiment, with local imperfections (Koschmieder 1974) [Dy, Fig. 142].

Figure 4.2: Dry salt lake.

Stationary convection can be 2-dimensional (convection rolls) or 3-dimensional (convection cells). Figures 4.1 and 4.2 give two examples of 3-dimensional convection: Fig. 4.1 shows an experiment in a lab, while Fig. 4.2 shows a dried salt lake in Africa. In the experiment we see clearly how an imperfection (little dip in the bottom) perturbs the regular hexagonals. The nature of the perturbation of the resulting steady state suggests that the solutions depend continuously on those quantities involved in the mathematical modeling and that the mathematical problem is well-posed (small changes in the data result in small changes of the flow). When we consider the dry salt lake we see that even more strongly perturbed configurations (uneven ground, evaporation of water, deposition of salt, ...) show qualitatively similar solutions.

Convective instability seems to have been first described by James Thomson (1882), a brother of Lord Kelvin. The first quantitative experiments were made by Bénard (1900)[2]. Stimulated by Bénard's experiments, Rayleigh (1916) formulated the theory of convective instability of a layer of fluid between horizontal planes. He showed that instability would occur only when the temperature difference was so large that the dimensionless parameter Ray defined in (4.9) exceeded a certain critical value. This parameter is now called the Rayleigh number. It is proportional to the magnitude of the temperature gradient if all other parameters are kept fixed.

In the following we shall be interested in 2-dimensional (2D) stationary states, and we shall use the Boussinesq approximation for their mathematical description. It should be stressed, however, that there are many situations in which the Boussinesq approximation does also describe 3-dimensional steady states and turbulent flows appropriately.

4.2.2 Boussinesq approximation

The Boussinesq approximation is widely used in the mathematical modeling of geophysical flows. It is a simplification of the basic equations of continuum mechanics and was introduced by Oberbeck (1879) and by Boussinesq (1903). It works well for flows with temperature differences of a few degrees or less and can be formally justified by dimensional analysis. The basis of this approximation is the observation that there are flows in which the temperature varies little, and thus the density varies little, yet in which the

[2]Bénard investigated the instability of a very thin layer of fluid with stress-free upper surface. As was shown around 1957, the instability of such layers is driven by variations of surface tension, not by variations of density [DR, p. 34].

buoyancy drives the motion. Then the variation of density is neglected everywhere except in the buoyancy term, where it cannot be ignored because the acceleration resulting from the external force in the equation of motion can be quite large, even larger than the acceleration due to the inertial term $(u \cdot \nabla)u$. A fluid for which the Boussinesq approximation is adequate is called **Boussinesq fluid**.

We consider a layer of fluid between two parallel planes of constant temperature; infinitely extended in the horizontal direction, i.e. boundary effects are neglected. Let

d be the thickness of the fluid layer,
ρ the density of the fluid at the upper plane,
κ its thermal diffusivity, and
ν its kinematic viscosity;
T the temperature at the upper plane, T_o the temperature at the lower plane, $T_o > T$,
$\Delta T := T_o - T > 0$, and
g the acceleration due to gravity.

The mathematical description of this situation starts with the basic equations from continuum mechanics for Newtonian fluids [Cha]. The following assumptions are made:

- The density ρ depends linearly on the temperature; its dependence on the pressure p is so small that it can be neglected:

$$\rho = \rho_o(1 - \alpha(T - T_o)), \tag{4.1}$$

  where α is a constant depending on the fluid to be considered. Typically, α is in the range 10^{-3} to 10^{-4} for variations in temperature not exceeding 10 degrees, say, and then $(\rho - \rho_o)/\rho_o = \alpha(T_o - T) \leq 10^{-2} \ll 1$, but nevertheless the buoyancy $g(\rho - \rho_o)$ is of the same order of magnitude as the inertia, acceleration or viscous stresses of the fluid and so is not negligible.
- The only external force $\mathbf{f}$ not to be neglected in the equation of motion is gravitation, which is a homogeneous field in the vertical direction:

$$f_i = -\rho g \delta_{i3}, \qquad i = 1, 2, 3. \tag{4.2}$$

- All material properties are assumed to be constant with the exception of the density in the gravity term.

In many cases these assumptions are justified by the size of the effects neglected. As was already explained in Chap. 1, it is a substantial feature of mathematical modeling to obtain as simple models as possible. All small effects that would make the equations more complicated than necessary for the investigation of the effects under interest should be neglected.

Thermodynamic relations are used to convert the equation for the internal energy e into an equation for the temperature T. The equations are made dimensionless by introducing d, d^2/κ and ΔT as scales for length, time and temperature, respectively. The reference speed then is κ/d.

We shall investigate stationary solutions bifurcating from the static solution

$$T = T_o - x_j\delta_{j3}, \qquad v_i = 0, \qquad i,j = 1,2,3, \tag{4.3}$$

where the v_i are the three velocity components and $x_j\delta_{j3} = x_3$ is the vertical space coordinate. Thus it is convenient to formulate the equations for stationary deviations from the static solution. Let θ be the temperature deviation from the static temperature distribution,

$$\theta := T - T_o + x_j\delta_{j3}. \tag{4.4}$$

Then the equations finally read [Cha, DR]:

$$\partial_j v_j = 0, \tag{4.5}$$

$$v_j\partial_j v_i = -\partial_i\varpi + \text{Ray Pr } \theta\delta_{i3} + \text{Pr}\Delta v_i \tag{4.6}$$

$$v_j\partial_j\theta = v_j\delta_{j3} + \Delta\theta \tag{4.7}$$

$$\varpi := p + \frac{d^3 g}{\kappa^2}x_j\delta_{j3} + \frac{1}{2}\text{Pr Ray } (x_j\delta_{j3})^2, \tag{4.8}$$

with

$$\text{Pr} := \frac{\nu}{\kappa}, \qquad \text{Ray} := \frac{\alpha g d^3\Delta T}{\nu\kappa}, \tag{4.9}$$

where α is the coefficient of thermal expansion introduced in (4.1), and the other quantities were defined at the beginning of this subsection.
Pr is called **Prandtl number** and is constant for a given fluid within this approximation. Ray is called **Rayleigh number**. It is a characteristic ratio of the destabilizing effect of buoyancy to the stabilizing effects of diffusion and dissipation. It is proportional to the magnitude of the vertical temperature gradient if the geometric configuration, the fluid and the temperature at one of the bounding planes are fixed.

As Chandrasekhar verified [Cha, §9, p. 18ff], no information is lost by using the Boussinesq approximation for the investigation of the thermal instability of the static layer: the perturbation equations obtained from the full equations of continuum mechanics are the same as the perturbation equations obtained from the Boussinesq approximation. However, the Boussinesq approximation provides the basis for further investigations in the nonlinear regime.

These equations have to be supplemented by boundary conditions. In most theoretical investigations of convection rolls periodic boundary conditions are used in the horizontal direction. This corresponds to neglecting end effects as we have done that for the Taylor problem as well. In the vertical direction, rigid or stress-free boundaries are used.

Rigid boundaries (no-slip boundary conditions, (4.50)–(4.55)) are adequate when the layer is contained between solid plates. We thus shall use these boundary conditions when we investigate the relation between the Taylor and Rayleigh-Bénard problems.

Free boundaries (i.e. boundary conditions with vanishing tangential stresses, (4.67)) are adequate when the layer is bounded by other fluids. The model problems which we shall use to study secondary bifurcations both have stress-free boundaries on top and bottom.

Another (lighter) fluid above the layer is easy to realize in an experiment and a common situation in nature. Another (heavier) fluid below the layer is not so common and would complicate experiments quite a bit. In most experiments, the layer is supported by a copper plate. In theoretical investigations, however, the case of stress-free boundaries above *and below* is treated quite often. Its main advantage is that the related equations are more convenient for mathematical treatment than those for the more realistic case with rigid support of the layer. In several cases, investigations were first performed for the case of stress-free boundaries on top and bottom, and later for the case with rigid boundaries. Since convection rolls have vanishing vertical vorticity and since they are the only stable solutions for small $\mathrm{Ray} - \mathrm{Ray}_{cr}$, results for both cases of boundary conditions are essentially the same in the regime of stable rolls. The extension of the theory of oscillations from free to rigid boundaries, however, does not merely involve quantitative aspects. The mechanism of the oscillatory instability is profoundly altered by the presence of rigid boundaries, since the occurrence of oscillations is connected with the appearance of vertical vorticity [CB74, p. 625f].

4.2.3 The Rayleigh-Bénard problem as limiting case of the Taylor problem

As mentioned earlier, Rayleigh already noticed that there is a general qualitative correspondence between the onset of Taylor vortex flows in narrow gaps between concentric cylinders and the onset of convection rolls between horizontal plates, and he used the analogy when deriving the stability criterion (1.1) for inviscid flows. Later on, many authors investigated the relation between the two problems and made use of it, see 'Narrow gap limit and Rayleigh-Bénard problem' of Chapter 1 and [Jef28, Cha, DR, Stu, Bue84, Bus, BO, Na86], for instance.

Nevertheless, Drazin, Reid were probably right when they stated in 1981 that 'an algorithm' was missing 'by which we can approximate the solution of the narrow gap Taylor problem asymptotically in terms of the Bénard problem with two rigid boundaries' [DR, p. 98].

Here we show how the limit $\eta \to 1, \quad \mu \to 1$ can be performed on the full 2D systems of equations in the primitive variables in such a way that it can be implemented into numerical codes. A single code is useful then for computing Taylor vortex flows in wide and narrow gaps *and* convection rolls; and it is possible to change gap width η and rotation rate μ simultaneously using pseudo-arclength continuation.

Chandrasekhar's observation is confirmed that the limits $\eta \to 1$ and $\mu \to 1$ cannot be performed independently, but have to be coordinated [Cha, eqs. (161), (162), and (191)]. We get the condition

$$\frac{1-\mu}{1-\eta} = 2\left(1 + \frac{\text{Ray}}{4\,\text{Re}^2}\right), \tag{4.10}$$

where $\text{Re} = \text{Re}^D$ is the Reynolds number as defined in (4.13) (Taylor problem) and Ray is the Rayleigh number as defined in (4.9) (Rayleigh-Bénard problem). The factors in formula (4.10) depend on the chosen quantities in the process of non-dimensionalizing (some of the popular ways of non-dimensionalizing the Taylor problem are forbidden, though, since they lead to terms which do not stay bounded in the limit). This limit is mathematical in nature: We have to identify the deviation of the toroidal velocity component v from Couette flow with the temperature deviation θ. The presentation follows [LMP].

It should be noted that both Taylor problem and Rayleigh-Bénard problem are limiting cases of the more general problem of rotating cylinders of different temperatures. This is a problem of interest by itself, and also because it generalizes both problems and unifies the theory for them [DR, KB91].

The Taylor problem

We consider again flows between two concentric rotating cylinders. As before, let R be the dimensional radius, let the inner cylinder have radius R_1 and angular velocity Ω_1, the outer cylinder radius R_2 and angular velocity Ω_2, and let

$$D = R_2 - R_1, \qquad \eta = R_1/R_2, \qquad \mu = \frac{\Omega_2}{\Omega_1}\,. \tag{4.11}$$

The gap width D is an appropriate length scale for the narrow gap limit, and $\Omega_1 D$ is an appropriate velocity. We thus define

$$r_1 := R_1/D, \qquad r_2 := R_2/D, \qquad \chi := \frac{R}{D} - r_1 \tag{4.12}$$

and

$$\mathrm{Re} := \mathrm{Re}^D := \frac{\Omega_1 D^2}{\nu}\,, \tag{4.13}$$

where ν is the kinematic viscosity. The Reynolds number Re^D used here deviates for $R_1 \neq D$ from $\mathrm{Re} = \Omega_1 R_1^2/\nu$ and from $\mathrm{Re}^\delta = \Omega_1 R_1 D/\nu$ used in the other chapters. Since we do not use these other definitions here, we do not expect confusion and thus use here the simpler Re in the formulas whenever Re^D occurs. The Navier-Stokes equations now read in non-dimensional form in cylindrical coordinates (see Fig. 4.3):

$$\partial_{\chi\chi}u + \frac{\partial_\chi u}{r_1+\chi} + \partial_{\varsigma\varsigma}u - \frac{u}{(r_1+\chi)^2} - \partial_\chi p = \mathrm{Re}\left(u\partial_\chi u + w\partial_\varsigma u - \frac{v^2}{r_1+\chi}\right), \tag{4.14}$$

$$\partial_{\chi\chi}v + \frac{\partial_\chi v}{r_1+\chi} + \partial_{\varsigma\varsigma}v - \frac{v}{(r_1+\chi)^2} = \mathrm{Re}\left(u\partial_\chi v + w\partial_\varsigma v + \frac{uv}{r_1+\chi}\right), \tag{4.15}$$

$$\partial_{\chi\chi}w + \frac{\partial_\chi w}{r_1+\chi} + \partial_{\varsigma\varsigma}w - \partial_\varsigma p = \mathrm{Re}(u\partial_\chi w + w\partial_\varsigma w), \tag{4.16}$$

$$\partial_\chi u + \frac{u}{r_1+\chi} + \partial_\varsigma w = 0. \tag{4.17}$$

We use the usual non-slip boundary conditions at the cylinders and periodic boundary conditions at the top and bottom and get

$$u(0,\zeta) = u(1,\zeta) = 0 \tag{4.18}$$

$$v(0,\zeta) = r_1, \qquad v(1,\zeta) = \mu(r_1+1) \tag{4.19}$$

$$w(0,\zeta) = w(1,\zeta) = 0 \tag{4.20}$$

for $-\pi/k \le \zeta \le \pi/k$ and

$$u(\chi,\frac{\pi}{k}) = u(\chi,-\frac{\pi}{k}) \tag{4.21}$$

$$v(\chi,\frac{\pi}{k}) = v(\chi,-\frac{\pi}{k}) \tag{4.22}$$

$$w(\chi,\frac{\pi}{k}) = w(\chi,-\frac{\pi}{k}) \tag{4.23}$$

for $0 \le \chi \le 1$. For small Reynolds number Re, this system has a unique solution, the Couette flow, which in our notation is given by

$$V(\chi) = a(r_1+\chi) + b/(r_1+\chi), \tag{4.24}$$

a and b being constants to be computed from the boundary conditions (4.19). The corresponding pressure P satisfies

$$\partial_\chi P = \frac{\mathrm{Re}\, V(\chi)^2}{r_1+\chi}. \tag{4.25}$$

With

$$\varepsilon := \frac{R_2-R_1}{R_2} = \frac{D}{R_2} = \frac{1}{r_2} = 1-\eta \tag{4.26}$$

and a quantity β satisfying

$$2\varepsilon(1+\beta) = 1-\mu \tag{4.27}$$

(β will be specified later), we get

$$a = 1-\frac{2(1+\beta)}{2-\varepsilon}, \qquad b = \frac{2(1+\beta)}{2-\varepsilon}\, r_1^2. \tag{4.28}$$

Equation (4.24) turns into

$$V(\chi) = \left(1-\frac{2(1+\beta)}{2-\varepsilon}\right)(r_1+\chi) + \frac{2(1+\beta)}{2-\varepsilon}\, r_1^2\, \frac{1}{r_1+\chi}. \tag{4.29}$$

We now use the expansions

$$\begin{aligned} v(\chi,\zeta) &= V(\chi)+v^*(\chi,\zeta), && (4.30)\\ p(\chi,\zeta) &= P(\chi)+p^*(\chi,\zeta) && (4.31)\end{aligned}$$

in eqs. (4.14)–(4.23) and get

$$\partial_{\chi\chi}u+\frac{1}{r_1+\chi}\partial_\chi u+\partial_{\zeta\zeta}u-\frac{u}{(r_1+\chi)^2}-\partial_\chi p^* \tag{4.32}$$
$$=\mathrm{Re}\left(u\partial_\chi u+w\partial_\zeta u-2v^*+\frac{4(1+\beta)v^*}{(2-\varepsilon)}-\frac{4(1+\beta)r_1^2v^*}{(2-\varepsilon)(r_1+\chi)^2}-\frac{v^{*2}}{r_1+\chi}\right),$$

$$\partial_{\chi\chi}v^*+\frac{1}{r_1+\chi}\,\partial_\chi v^*+\partial_{\zeta\zeta}v^*-\frac{v^*}{(r_1+\chi)^2} \tag{4.33}$$
$$=\mathrm{Re}\left(u\partial_\chi v^*+w\partial_\zeta v^*+2u-\frac{4(1+\beta)u}{2-\varepsilon}+\frac{uv^*}{r_1+\chi}\right),$$

$$\partial_{\chi\chi}w+\frac{1}{r_1+\chi}\partial_\chi w+\partial_{\zeta\zeta}w-\partial_\zeta p^*=\mathrm{Re}(u\partial_\chi w+w\partial_\zeta w), \tag{4.34}$$

$$\partial_\chi u+\frac{u}{r_1+\chi}+\partial_\zeta w=0. \tag{4.35}$$

We now consider $R_1,R_2\to\infty$, i.e. $\varepsilon\to 0$, and assume that all functions involved stay bounded. Thus (4.32)–(4.35) become

$$\begin{aligned} \partial_{\chi\chi}u+\partial_{\zeta\zeta}u-\partial_\chi p^* &= \mathrm{Re}(u\partial_\chi u+w\partial_\zeta u-2v^*), && (4.36)\\ \partial_{\chi\chi}v^*+\partial_{\zeta\zeta}v^* &= \mathrm{Re}(u\partial_\chi v^*+w\partial_\zeta v^*-2\beta u), && (4.37)\\ \partial_{\chi\chi}w+\partial_{\zeta\zeta}w-\partial_\zeta p^* &= \mathrm{Re}(u\partial_\chi w+w\partial_\zeta w), && (4.38)\\ \partial_\chi u+\partial_\zeta w &= 0, && (4.39)\end{aligned}$$

with the boundary conditions

$$\begin{aligned} u(0,\zeta) &= u(1,\zeta)=0 && (4.40)\\ v^*(0,\zeta) &= v^*(1,\zeta)=0 && (4.41)\\ w(0,\zeta) &= w(1,\zeta)=0 && (4.42)\end{aligned}$$

for $-\pi/k\le\zeta\le\pi/k$, and

$$\begin{aligned} u(\chi,\frac{\pi}{k}) &= u(\chi,-\frac{\pi}{k}) && (4.43)\\ v^*(\chi,\frac{\pi}{k} &= v^*(\chi,-\frac{\pi}{k}) && (4.44)\\ w(\chi,\frac{\pi}{k} &= w(\chi,-\frac{\pi}{k}) && (4.45)\end{aligned}$$

for $0\le\chi\le 1$.

Identification of the equations

We now investigate the relation between system (4.36)–(4.45) and the equations of the Boussinesq approximation, eqs. (4.5)–(4.9):

$$
\begin{aligned}
(v_x\partial_x + v_z\partial_z)v_x &= -\partial_x\varpi + \mathrm{Pr}\ \Delta v_x, &(4.46)\\
(v_x\partial_x + v_z\partial_z)v_z &= -\partial_x\varpi + \mathrm{Pr\ Ray}\ \theta + \mathrm{Pr}\ \Delta v_z, &(4.47)\\
(v_x\partial_x + v_z\partial_z)\theta &= v_z + \Delta\theta, &(4.48)\\
\partial_x v_x + \partial_z v_z &= 0, &(4.49)
\end{aligned}
$$

with

$$
\begin{aligned}
v_x(x,0) &= v_x(x,1) = 0 &(4.50)\\
v_z(x,0) &= v_z(x,1) = 0 &(4.51)\\
\theta(x,0) &= \theta(x,1) = 0 &(4.52)
\end{aligned}
$$

for $-\alpha/2 \le x \le \alpha/2$ and

$$
\begin{aligned}
v_x(-\frac{\alpha}{2},z) &= v_x(\frac{\alpha}{2},z) &(4.53)\\
v_z(-\frac{\alpha}{2},z) &= v_z(\frac{\alpha}{2},z) &(4.54)\\
\theta(-\frac{\alpha}{2},z) &= \theta(\frac{\alpha}{2},z) &(4.55)
\end{aligned}
$$

for $0 \le z \le 1$. The geometrical configurations are illustrated in Fig. 4.3. They suggest that we put

$$
\begin{aligned}
\chi &= z, \qquad \zeta = -x &(4.56)\\
u &= \alpha_2 v_z, \qquad v^* = \gamma\theta, \qquad w = -\alpha_1 v_x, &(4.57)\\
p^* &= \delta\varpi. &(4.58)
\end{aligned}
$$

The α_i, γ, δ are to be specified. Comparison of the two versions of the equation of continuity (4.39) and (4.49) yields

$$\alpha_1 = \alpha_2 =: \alpha. \tag{4.59}$$

Equations (4.36), (4.47), (4.56) and (4.59) now yield

$$\alpha = 1/\mathrm{RePr}, \qquad 2\gamma\mathrm{Re} = \alpha\mathrm{Ray}, \qquad \delta = \alpha/\mathrm{Pr}. \tag{4.60}$$

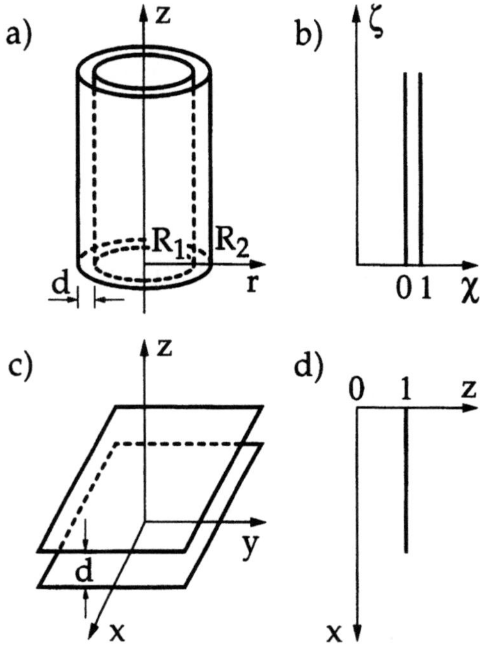

Figure 4.3: Geometrical configurations and coordinate systems
a), b): Taylor problem; c), d): Rayleigh-Bénard problem [LMP, Fig. 4].

Equation (4.37) for the toroidal velocity deviation v and eq. (4.48) for the temperature deviation θ now yield

$$\alpha \mathrm{Re} = 1, \qquad 2\alpha\beta \mathrm{Re} = \gamma. \tag{4.61}$$

Equation (4.38) for the axial velocity component w and equation (4.46) for the horizontal velocity v_x now yield

$$\delta = \alpha/\mathrm{Pr}, \qquad \alpha = 1/\mathrm{PrRe}. \tag{4.62}$$

We now solve system (4.56)–(4.62) and get

$$\begin{aligned} \mathrm{Pr} &= 1, \qquad \alpha = \alpha_1 = \alpha_2 = 1/\mathrm{Re}, \\ \beta &= \mathrm{Ray}/4\mathrm{Re}^2, \qquad \gamma = \mathrm{Ray}/2\mathrm{Re}^2, \\ \delta &= 1/\mathrm{Re}, \end{aligned} \tag{4.63}$$

and (4.56) now reads

$$\begin{aligned} \chi &= z, & \zeta &= -x, \\ \text{Re}\ u &= v_z, & 2\ \text{Re}^2\ v^* &= \text{Ray}\ \theta, \\ \text{Re}\ w &= -v_x, & \text{Re}\ p^* &= \varpi. \end{aligned} \tag{4.64}$$

The boundary conditions (4.40)–(4.45) and (4.50)–(4.55) do not cause any problems. Also, they do not give any additional information.

We now have a closer look at eqs. (4.26)–(4.29). Using eq. (4.63), we now get from (4.27)

$$2\left(1+\frac{\text{Ray}}{4\text{Re}^2}\right) = \frac{1-\mu}{1-\eta}. \tag{4.65}$$

Since the left side of (4.65) is finite for all Ray and Re, this must be true for the right side as well: μ and η cannot attain their limits independently, $(1-\mu)/(1-\eta)$ has to stay bounded.

4.3 Stationary convection rolls

4.3.1 The basic equations

Let x and z denote the horizontal and the vertical direction, respectively. The velocity is denoted by v, and the temperature deviation from the conductive solution by θ. Using the Boussinesq approximation, convection rolls are described by eqs. (4.5)– (4.7):

$$\begin{aligned} (v_x\partial_x + v_z\partial_z)v_x &= -\partial_x\varpi + \text{Pr}\ \Delta v_x, \\ (v_x\partial_x + v_z\partial_z)v_z &= -\partial_x\varpi + \text{Pr Ray}\ \theta + \text{Pr}\ \Delta v_z, \\ (v_x\partial_x + v_z\partial_z)\theta &= v_z + \Delta\theta, \\ \partial_x v_x + \partial_z v_z &= 0. \end{aligned} \tag{4.66}$$

We are going to study the perfect problem with stress-free boundary conditions (no tangential stresses) in the vertical direction and periodic boundary conditions with period $2\pi/\alpha$ in the horizontal direction. The boundary conditions then read

$$\begin{aligned} \partial_z v_x(x,0) &= \partial_z v_x(x,1) = 0, \\ v_z(x,0) &= v_z(x,1) = 0, \qquad 0 \le x \le \frac{2\pi}{\alpha} \\ \theta(x,0) &= \theta(x,1) = 0, \end{aligned} \tag{4.67}$$

$$\begin{aligned} v_x(0,z) &= v_x(\frac{2\pi}{\alpha},z), \\ v_z(0,z) &= v_z(\frac{2\pi}{\alpha},z), \qquad\qquad 0 \le z \le 1, \\ \theta(0,z) &= \theta(\frac{2\pi}{\alpha},z). \end{aligned}$$

The configuration and the equations allow a reflection symmetry with respect to the mid-plane $z \equiv 1/2$ which we can formulate for the vertical velocity v_z as [BO]

$$v_z(x, z-\frac{1}{2}) = -v_z(\frac{\pi}{\alpha} - x, -z+\frac{1}{2}). \tag{4.68}$$

We introduce the stream function ψ satisfying $v_x = -\partial_z\psi, \quad v_z = \partial_x\psi$, eliminate ϖ and get from eqs. (4.66)–(4.67)

$$\begin{aligned} (\partial_z\psi\partial_x - \partial_x\psi\partial_z)\Delta\psi &= -\text{Pr Ray } \partial_x\theta - \text{Pr } \Delta^2\psi, & (4.69) \\ (\partial_z\psi\partial_x - \partial_x\psi\partial_z)\theta &= -\partial_x\psi - \Delta\theta & (4.70) \end{aligned}$$

with the boundary conditions

$$\begin{aligned} \partial_{zz}\psi(x,0) &= \partial_{zz}\psi(x,1) = 0, \\ \psi(x,0) &= \psi(x,1) = 0, \qquad\qquad 0 \le x \le \frac{2\pi}{\alpha}, \\ \theta(x,0) &= \theta(x,1) = 0, \qquad\qquad\qquad\qquad (4.71) \\ \partial_z\psi(0,z) &= \partial_z\psi(\frac{2\pi}{\alpha},z), \\ \partial_x\psi(0,z) &= \partial_x\psi(\frac{2\pi}{\alpha},z), \qquad\qquad 0 \le z \le 1, \\ \theta(0,z) &= \theta(\frac{2\pi}{\alpha},z). \end{aligned}$$

This suggests a Fourier expansion in x and z, which we write as

$$\psi(x,z) = \sum_{i=1}^{M}\sum_{j=1}^{N} \psi_{ij}\sin(i\alpha x)\sin(j\pi z), \tag{4.72}$$

$$\theta(x,z) = \sum_{k=0}^{M}\sum_{l=1}^{N} \theta_{kl}\cos(k\alpha x)\sin(l\pi z). \tag{4.73}$$

This ansatz restricts our solution manifold to solutions which are skew-symmetric with respect to x. According to Busse/Or [BO], we are actually losing solutions that way. As explained in the introduction, we are mostly interested in curves similar to the curve of secondary bifurcations shown in Fig. 3.24. The related solutions are skew-symmetric in the horizontal direction. This justifies the approach (4.72). As is seen from eq. (4.70), ψ and θ are phase-shifted by $\pi/2$ in the x-direction, and this explains the ansatz (4.73). With the ansatz (4.72), (4.73), the boundary conditions (4.71) are identically satisfied.

Now we insert eqs. (4.72), (4.73) into eqs. (4.69), (4.70), multiply eq. (4.69) by $\sin(p\alpha x)\sin(q\pi z)$ and eq. (4.70) by $\cos(p\alpha x)\sin(q\pi z)$, $p \geq 0$, $q > 0$, use the addition formulas for products of trigonometric functions, and integrate. This gives us

$$\sum_{i,m=1}^{M}\sum_{j,n=1}^{N} \psi_{ij}\psi_{mn}a_{mn}\Big[b_{in}(\delta_{i,|p-m|}-\delta_{i,p+m})(\delta_{j,q+n}+sign(q-n)\delta_{j,|q-n|})+ \\ -b_{mj}(\delta_{i,p+m}+sign(p-m)\delta_{i,|p-m|})(\delta_{j,|q-n|}-\delta_{j,q+n})\Big] \\ = \text{Pr Ray } c_p\theta_{pq} - \text{Pr } a_{pq}^2\psi_{pq}; \tag{4.74}$$

$$\sum_{i,k=1}^{M}\sum_{j,\ell=1}^{N} \psi_{ij}\theta_{k\ell}\Big[b_{kj}(sign(k-p)\delta_{i,|p-k|}+\delta_{i,p+k})(\delta_{j,|\ell-q|}-\delta_{j,\ell+q})+ \\ +b_{i\ell}(\delta_{i,p+k}+\delta_{i,|p-k|})(sign(q-\ell)\delta_{j,|q-\ell|}+\delta_{j,q+\ell})\Big] \\ = c_p\psi_{pq} - a_{pq}\theta_{pq} \quad \text{for } p \neq 0; \tag{4.75}$$

$$\sum_{i=0}^{M}\sum_{j,\ell=1}^{N} \psi_{ij}\theta_{i\ell}\Big[(b_{ij}+sign(q-\ell)b_{i\ell})\delta_{j,|\ell-q|}-(b_{ij}-b_{i\ell})\delta_{j,\ell+q}\Big] \\ = -a_{0q}\theta_{0q}, \tag{4.76}$$

with the abbreviations

$$a_{ij} := i^2\alpha^2 + j^2\pi^2, \qquad b_{ij} := \frac{i\alpha j\pi}{4}, \qquad c_i := i\alpha. \tag{4.77}$$

We thus obtain an algebraic system of $MN + (M+1)N = N(2M+1)$ equations. For sufficiently large M and N this is a satisfactory Galerkin approximation to eqs. (4.69)–(4.71), see the corresponding discussion of the Galerkin approach for the Taylor problem in Chapter 2. In following sections, however, we shall use very small values of N and M, i.e. we shall consider *model problems* obtained through severe truncation of the series ($N = 2,\ M \leq 3$). We shall not perform convergence studies as we did in

Chapter 2, but we shall compare the results obtained for a model problem to the results obtained for a perturbation approach to $3rd$ order. Before that, we shall discuss some results that are valid for system (4.74)–(4.76) for large N and M.

Though we are interested in results for large N and M at present, it is convenient to consider those 14 equations which we obtain by truncating system (4.74)–(4.76) at $N = 2, \quad M = 3$, and use higher values of N and M where needed. We thus get

$$\begin{aligned}(b_{11} - b_{22})(a_{21} - a_{12})\psi_{12}\psi_{21} + (b_{21} - b_{32})(a_{31} - a_{22})\psi_{22}\psi_{31} \\ + (b_{31} - b_{22})(a_{32} - a_{21})\psi_{21}\psi_{32} \\ = \text{Pr Ray } c_1\theta_{11} - \text{Pr } a_{11}^2\psi_{11},\end{aligned} \tag{4.78}$$

$$\begin{aligned}(b_{11} + b_{12})(a_{11} - a_{12})\psi_{11}\psi_{12} + (b_{11} - b_{32})(a_{31} - a_{12})\psi_{12}\psi_{31} \\ + (b_{12} - b_{31})(a_{11} - a_{32})\psi_{11}\psi_{32} \\ = \text{Pr Ray } c_2\theta_{21} - \text{Pr } a_{21}^2\psi_{21},\end{aligned} \tag{4.79}$$

$$\begin{aligned}(b_{12} + b_{21})(a_{11} - a_{22})\psi_{11}\psi_{22} + (b_{11} + b_{22})(a_{21} - a_{12})\psi_{21}\psi_{12} \\ = \text{Pr Ray } c_3\theta_{31} - \text{Pr } a_{31}^2\psi_{31},\end{aligned} \tag{4.80}$$

$$\begin{aligned}(b_{11} + b_{21})(a_{21} - a_{11})\psi_{11}\psi_{21} + (b_{21} + b_{31})(a_{31} - a_{21})\psi_{21}\psi_{31} \\ = \text{Pr Ray } c_1\theta_{12} - \text{Pr } a_{12}^2\psi_{12},\end{aligned} \tag{4.81}$$

$$\begin{aligned}(b_{11} + b_{31})(a_{31} - a_{11})\psi_{11}\psi_{31} \\ = \text{Pr Ray } c_2\theta_{22} - \text{Pr } a_{22}^2\psi_{22},\end{aligned} \tag{4.82}$$

$$\begin{aligned}(b_{11} - b_{21})(a_{21} - a_{11})\psi_{21}\psi_{11} \\ = \text{Pr Ray } c_3\theta_{32} - \text{Pr } a_{32}^2\psi_{32},\end{aligned} \tag{4.83}$$

$$\begin{aligned}-2b_{12}\psi_{11}\theta_{02} + 2b_{11}\psi_{12}\theta_{01} + (b_{11} - b_{22})(\psi_{21}\theta_{12} + \psi_{12}\theta_{21}) \\ + (b_{21} - b_{32})(\psi_{31}\theta_{22} + \psi_{22}\theta_{31}) \\ - (b_{22} - b_{31})(\psi_{32}\theta_{21} + \psi_{21}\theta_{32}) \\ = c_1\psi_{11} - a_{11}\theta_{11},\end{aligned} \tag{4.84}$$

$$\begin{aligned}-2b_{22}\psi_{21}\theta_{02} + 2b_{21}\psi_{22}\theta_{01} + (b_{11} - b_{32})(\psi_{31}\theta_{12} + \psi_{12}\theta_{31}) \\ - (b_{12} - b_{31})(\psi_{32}\theta_{11} + \psi_{11}\theta_{32}) \\ - (b_{11} + b_{12})(\psi_{11}\theta_{12} - \psi_{12}\theta_{11}) \\ = c_2\psi_{21} - a_{21}\theta_{21},\end{aligned} \tag{4.85}$$

$$\begin{aligned}
-2b_{32}\psi_{31}\theta_{02} + 2b_{31}\psi_{32}\theta_{01} - (b_{11}+b_{22})(\psi_{21}\theta_{12} - \psi_{12}\theta_{21}) \\
+ (b_{12}+b_{21})(\psi_{22}\theta_{11} - \psi_{11}\theta_{22}) \\
= c_3\psi_{31} - a_{31}\theta_{31},
\end{aligned} \tag{4.86}$$

$$\begin{aligned}
2b_{11}\psi_{11}\theta_{01} + (b_{11}+b_{21})(\psi_{21}\theta_{11} + \psi_{11}\theta_{21}) \\
+ (b_{21}+b_{31})(\psi_{31}\theta_{21} + \psi_{21}\theta_{31}) \\
= c_1\psi_{12} - a_{12}\theta_{12},
\end{aligned} \tag{4.87}$$

$$\begin{aligned}
2b_{21}\psi_{21}\theta_{01} + (b_{11}+b_{31})(\psi_{31}\theta_{11} + \psi_{11}\theta_{31}) \\
= c_2\psi_{22} - a_{22}\theta_{22},
\end{aligned} \tag{4.88}$$

$$\begin{aligned}
2b_{31}\psi_{31}\theta_{01} + (b_{11}-b_{21})(\psi_{11}\theta_{21} - \psi_{21}\theta_{11}) \\
= c_3\psi_{32} - a_{32}\theta_{32},
\end{aligned} \tag{4.89}$$

$$\begin{aligned}
(b_{11}-b_{12})\cdot(\psi_{11}\theta_{12} + \psi_{12}\theta_{11}) + (b_{21}-b_{22})(\psi_{21}\theta_{22} + \psi_{22}\theta_{21}) + \\
+ (b_{31}-b_{32})(\psi_{31}\theta_{32} + \psi_{32}\theta_{31}) \\
= -\, a_{01}\theta_{01},
\end{aligned} \tag{4.90}$$

$$\begin{aligned}
2b_{11}\psi_{11}\theta_{11} + 2b_{21}\psi_{21}\theta_{21} + 2b_{31}\psi_{31}\theta_{31} \\
= -\, a_{02}\theta_{02},
\end{aligned} \tag{4.91}$$

4.3.2 Critical curves of the primary solution

System (4.78)–(4.91) depends on three parameters: Rayleigh number Ray, Prandtl number Pr and the wave number α, which is hidden in the coefficients defined in (4.77). For small values of Ray the system is uniquely solvable for all values of Pr and α. For larger Ray, this *conduction solution* $\Psi = 0,\ \Theta = 0$ is unstable, and convection rolls occur. To compute the critical values for the bifurcation of the convection solutions from the conduction solution, we linearize system (4.78)–(4.91) at the conduction solution and obtain

$$\psi_{11}a_{11}^2 - \text{Ray}\ \theta_{11}c_1 = 0, \tag{4.92}$$
$$\psi_{12}a_{12}^2 - \text{Ray}\ \theta_{12}c_1 = 0, \tag{4.93}$$
$$\psi_{21}a_{21}^2 - \text{Ray}\ \theta_{21}c_2 = 0, \tag{4.94}$$
$$\psi_{22}a_{22}^2 - \text{Ray}\ \theta_{22}c_2 = 0, \tag{4.95}$$

$$\begin{aligned}
\psi_{31}a_{31}^2 - \text{Ray}\ \theta_{31}c_3 &= 0, && (4.96)\\
\psi_{32}a_{32}^2 - \text{Ray}\ \theta_{32}c_3 &= 0, && (4.97)\\
\theta_{01} &= 0, && (4.98)\\
\theta_{02} &= 0, && (4.99)\\
\psi_{11}c_1 - \theta_{11}a_{11} &= 0, && (4.100)\\
\psi_{12}c_1 - \theta_{12}a_{12} &= 0, && (4.101)\\
\psi_{21}c_2 - \theta_{21}a_{21} &= 0, && (4.102)\\
\psi_{22}c_2 + \theta_{22}a_{22} &= 0. && (4.103)\\
\psi_{31}c_3 - \theta_{31}a_{31} &= 0, && (4.104)\\
\psi_{32}c_3 + \theta_{32}a_{32} &= 0. && (4.105)
\end{aligned}$$

System (4.92)–(4.105) depends only on Ray and α, but not on Pr, and it is reducible. The quantities ψ_{11} and θ_{11} occur only in two of the equations, in (4.92) and (4.100). We thus get the critical values $\text{Ray}_{cr}^{11}(\alpha)$ for the lowest modes ψ_{11}, θ_{11} from a 2×2 determinant. The critical curve reads

$$R_{cr}^{11}(\alpha) := \frac{a_{11}^3}{c_1^2} = \frac{(\alpha^2 + \pi^2)^3}{\alpha^2}. \tag{4.106}$$

It gives the loci in the (α, Ray)-plane where the 2-roll convection solutions bifurcate from the conduction solution.

Along the same lines we get the critical curves for the other modes ψ_{pq}, θ_{pq} since these formulas are independent of the truncation parameters N and M, as long as $p < M,\ q < N$. They thus are the same for the linearization of system (4.74)–(4.76) for large $M,\ N$, and the determinant reads in general

$$\begin{vmatrix} a_{pq}^2 & -c_p\text{Ray} \\ c_p & -a_{pq} \end{vmatrix} \tag{4.107}$$

and leads to

$$R_{cr}^{pq}(\alpha) = \frac{a_{pq}^3}{c_p^2} = \frac{(p^2\alpha^2 + q^2\pi^2)^3}{(p\alpha)^2} \tag{4.108}$$

for the p, q^{th} critical curve. The first 6 curves ($p \leq 3,\ q \leq 2$) are shown in Fig. 4.4.

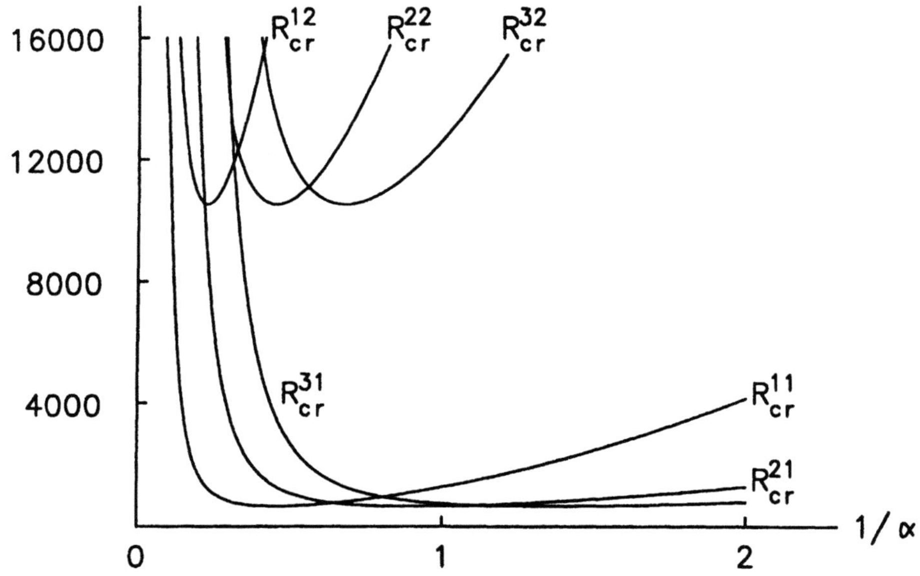

Figure 4.4: First six critical curves $R^{pq}_{cr}(\alpha)$ as given in eq. (4.108), plotted vs. α^{-1}.

Several of these curves intersect. Some α-values of intersection points are

$$\alpha^{11}_{21} = \pi\left(\frac{4^{1/3}-1}{4-4^{1/3}}\right)^{1/2}; \quad \alpha^{12}_{21} = \pi\left(\frac{4\cdot 4^{1/3}-1}{4-4^{1/3}}\right)^{1/2};$$

$$\alpha^{12}_{22} = \pi\left(\frac{4\cdot 4^{1/3}-4}{4-4^{1/3}}\right)^{1/2} = 2\alpha^{11}_{21}; \qquad (4.109)$$

$$\alpha^{11}_{31} = \pi\left(\frac{9^{1/3}-1}{9-9^{1/3}}\right)^{1/2}; \quad \alpha^{22}_{31} = \pi\left(\frac{4\cdot 9^{1/3}-4^{1/3}}{9\cdot 4^{1/3}-4\cdot 9^{1/3}}\right)^{1/2}.$$

In general, we get from $R^{ij}_{cr}(\alpha) = R^{mn}_{cr}(\alpha)$ the expression

$$\alpha^{ij}_{mn} = \pi\left(\frac{i^{2/3}\cdot n^2 - j^2\cdot m^{2/3}}{m^{2/3}\cdot i^2 - m^2\cdot i^{2/3}}\right)^{1/2}, \qquad i \neq m. \qquad (4.110)$$

For $i = m,\ j \neq n$ the curves do not intersect. Also, they do not intersect if (4.110) produces non-real complex numbers, as for instance $\alpha^{11}_{22} = \pi\sqrt{-1}$.

4.3.3 Pure-mode solutions

To find pure-mode 2-roll solutions of system (4.74)–(4.76) or of system (4.78)–(4.91), we first assume the coefficients of all modes to vanish, except ψ_{11} and θ_{11}. Under these assumptions, three equations of system (4.78)–(4.91) have to be considered. Eqs. (4.78) and (4.84) together lead to the relation (4.106) for $R_{cr}^{11}(\alpha)$ and thus to $\psi_{11} = \theta_{11} = 0$. This is consistent with the requirements of eq. (4.91).

If we assume ψ_{11}, θ_{11} and θ_{02} to be non-zero, eq. (4.84) does contain an additional term that allows the solutions to exist for other (α, Ray) values than those on the critical curve. Also, eq. (4.91) does allow non-vanishing solutions now. A non-trivial pure-mode solution hence does exist. It is given by

$$\begin{aligned}
\psi_{11}^2(\alpha, \mathrm{Ray}) &= \frac{8}{a_{11}^2(\alpha)}\left(\mathrm{Ray} - R_{cr}^{11}(\alpha)\right) = \frac{a_{02}c_1^2}{4b_{11}b_{12}a_{11}^2}\left(\mathrm{Ray} - R_{cr}^{11}\right),\\
\theta_{11}^2(\alpha, \mathrm{Ray}) &= \frac{8a_{11}^2(\alpha)}{\alpha^2}\,\frac{\mathrm{Ray} - R_{cr}^{11}(\alpha)}{\mathrm{Ray}^2} = \frac{a_{02}a_{11}^2}{4b_{11}b_{12}}\,\frac{\mathrm{Ray} - R_{cr}^{11}}{\mathrm{Ray}^2},\\
\theta_{02}(\alpha, \mathrm{Ray}) &= -\frac{\mathrm{Ray} - R_{cr}^{11}(\alpha)}{\pi\,\mathrm{Ray}} = -\frac{c_1}{2b_{12}}\,\frac{\mathrm{Ray} - R_{cr}^{11}}{\mathrm{Ray}}.
\end{aligned} \tag{4.111}$$

Similarly we assume $\psi_{21}, \theta_{21}, \theta_{02} \neq 0$ and compute the Fourier coefficients of the 4-roll solutions:

$$\begin{aligned}
\psi_{21}^2(\alpha, \mathrm{Ray}) &= \frac{8}{a_{21}^2(\alpha)}\left(\mathrm{Ray} - R_{cr}^{21}(\alpha)\right) = \frac{a_{02}c_2^2}{4b_{21}b_{22}a_{21}^2}\left(\mathrm{Ray} - R_{cr}^{21}\right)\\
\theta_{21}^2(\alpha, \mathrm{Ray}) &= \frac{2a_{21}^2(\alpha)}{\alpha^2}\,\frac{\mathrm{Ray} - R_{cr}^{21}(\alpha)}{\mathrm{Ray}^2} = \frac{a_{02}a_{21}^2}{4b_{21}b_{22}}\,\frac{\mathrm{Ray} - R_{cr}^{21}}{\mathrm{Ray}^2},\\
\theta_{02}(\alpha, \mathrm{Ray}) &= -\frac{\mathrm{Ray} - R_{cr}^{21}(\alpha)}{\pi\mathrm{Ray}} = -\frac{c_2}{2b_{22}}\,\frac{\mathrm{Ray} - R_{cr}^{21}}{\mathrm{Ray}}.
\end{aligned} \tag{4.112}$$

Assuming that only $\psi_{p1}, \theta_{p1}, \theta_{02} \neq 0$ we get similarly

$$\begin{aligned}
\psi_{p1}^2 &= \frac{a_{02}c_p^2}{4b_{p1}b_{p2}a_{p1}^2}\left(\mathrm{Ray} - R_{cr}^{p1}\right)\\
\theta_{p1}^2 &= \frac{a_{02}a_{p1}^2}{4b_{p1}b_{p2}}\,\frac{\mathrm{Ray} - R_{cr}^{p1}}{\mathrm{Ray}^2},\\
\theta_{02} &= -\frac{c_p}{2b_{p2}}\,\frac{\mathrm{Ray} - R_{cr}^{p1}}{\mathrm{Ray}}.
\end{aligned} \tag{4.113}$$

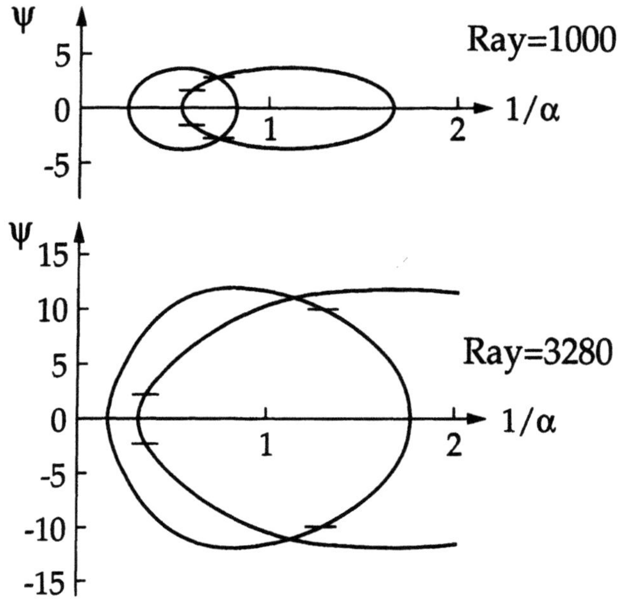

Figure 4.5: The pure-mode solutions ψ_{11} and ψ_{21} of eqs. (4.111) and (4.112) for the fixed Ray values Ray $= 1000$ and Ray $= 3280$. Also shown are the loci of secondary bifurcations to be discussed in the following, for Pr $= 1000$.

The solutions ψ_{11} and ψ_{21} are shown in Fig. 4.5 as functions of α^{-1} for fixed Ray values.

The solutions (4.111) and (4.112) were computed by us [LMP] for the 7×7 model problem (4.114). Here we used the assumption that only $\psi_{p1}, \theta_{p1}, \theta_{02} \neq 0$, $p \geq 1$ fixed, but did not truncate the system otherwise. We note that the periodicity of the solutions of the full system (4.74)–(4.76) is preserved by the solutions given in (4.111)–(4.113): $\psi_{21}(\alpha^{-1}) = \psi_{11}(2\alpha^{-1})$, $\psi_{p1}(\alpha^{-1}) = \psi_{11}(p\alpha^{-1})$, etc. This is seen from definitions (4.77) of the coefficients and from the periodicity behavior of the critical curves (4.108).

If we now try to find similarly pure-mode solutions bifurcating at the critical curves $\mathrm{Ray}_{cr}^{p2}(\alpha)$ from the trivial solution, we do not succeed:

The assumption that only ψ_{12} and θ_{12} do not vanish again leads to the formula of the critical curve and to $\psi_{12} = \theta_{12} = 0$. Equation (4.90) is consistent to that.

The assumption that only ψ_{12}, θ_{12} and θ_{01} do not vanish, however, does not lead to a pure-mode solution: eq. (4.90) gives a different mix of indices than eq. (4.91), and this enforces ψ_{11}, θ_{11} and θ_{02} to be non-zero for a non-trivial solution. Equation (4.85) then enforces $\psi_{21} \neq 0$, and so on. Definitely no pure-mode solutions.

Also, we do not see any other assumption that could lead to pure-mode representations of the double roll solutions. We thus conclude that the solutions bifurcating at the curves $\mathrm{Ray}_{cr}^{p2}(\alpha)$ are always mixed-mode solutions.

4.4 The (2,4) interaction in a model problem

System (4.74)–(4.76) and also system (4.78)–(4.91) are too complicated to be solved explicitly for the loci of secondary bifurcation points on the solutions just computed. Since we want to investigate the hypothesis that the nonlinear interaction of pure-mode solutions of different numbers of rolls generates curves of secondary bifurcations connecting intersection points of critical curves, the first question is: is there *any* system with such a curve of secondary bifurcations? This question was studied analytically by Lortz, Meyer-Spasche, Petroff [Pe88, LMP] for the interaction of the pure-mode 2-roll solution with the pure-mode 4-roll solution in a 7×7 *model problem.* Matern and Meyer-Spasche [Mat90] re-did these calculations and then investigated similarly the interaction of the pure-mode 2-roll solution with the pure-mode 6-roll solution in a different 7×7 model problem. The $(2,4)$-interaction was found to behave as expected from the results discussed in Chap. 3 (near Fig. 3.24). We review here those findings and study related problems.

4.4.1 The model problem

Truncating system (4.74)–(4.76) by a Galerkin approach with $N = M = 2$ gives us a 10×10 system. We get it from system (4.78)–(4.91) by replacing $\psi_{3q},\ \theta_{3q},\ \ q = 1, 2$ by zero, and by omitting eqs. (4.80), (4.83), (4.86) and (4.89). This system is now further reduced using perturbation arguments: Assuming

$$\psi_{11}, \psi_{21}, \theta_{11}, \theta_{21} \sim \varepsilon$$

(ε small), we conclude from (4.91) that $\theta_{02} \sim \varepsilon^2$. From eqs. (4.81) and (4.87) we conclude $\psi_{12}, \theta_{12} \sim \varepsilon^2$. Under the additional assumption $\theta_{01} \sim \varepsilon^3$ we conclude from eq.(4.88) that $\psi_{22}, \theta_{22} \sim \varepsilon^4$. This is consistent with eq. (4.90). It now follows that $\psi_{11}\theta_{01} \sim \varepsilon^4$, $\quad \psi_{12}\theta_{01} \sim \varepsilon^5$ and $\psi_{22}\theta_{01} \sim \varepsilon^7$.

We now neglect eqs. (4.82), (4.90) and (4.88) and all terms of orders higher/equal ε^3 in the remaining equations. With these simplifications, we obtain a 7×7 system which reads

$$
\begin{aligned}
(b_{11}-b_{22})(a_{21}-a_{12})\psi_{12}\psi_{21} - \text{Pr Ray } c_1\theta_{11} + \text{Pr } a_{11}^2\psi_{11} &= 0,\\
(b_{11}+b_{12})(a_{11}-a_{12})\psi_{11}\psi_{12} - \text{Pr Ray } c_2\theta_{21} + \text{Pr } a_{21}^2\psi_{21} &= 0,\\
(b_{11}+b_{21})(a_{21}-a_{11})\psi_{11}\psi_{21} - \text{Pr Ray } c_1\theta_{12} + \text{Pr } a_{12}^2\psi_{12} &= 0,\\
-2b_{12}\psi_{11}\theta_{02} + (b_{11}-b_{22})(\psi_{21}\theta_{12}+\psi_{12}\theta_{21}) - c_1\psi_{11} + a_{11}\theta_{11} &= 0,\\
-2b_{22}\psi_{21}\theta_{02} - (b_{11}+b_{12})(\psi_{11}\theta_{12}-\psi_{12}\theta_{11}) - c_2\psi_{21} + a_{21}\theta_{21} &= 0,\\
(b_{11}+b_{21})(\psi_{21}\theta_{11}+\psi_{11}\theta_{21}) - c_1\psi_{12} + a_{12}\theta_{12} &= 0,\\
2b_{11}\psi_{11}\theta_{11} + 2b_{21}\psi_{21}\theta_{21} + a_{02}\theta_{02} &= 0.
\end{aligned}
\tag{4.114}
$$

4.4.2 Calculation of secondary bifurcation points on the 2-roll solutions

A necessary condition for secondary bifurcation is that the functional determinant of system (4.114) vanishes on the solution branch under consideration. This means that the following determinant has to vanish:

$$
\begin{vmatrix}
-a_{02} & -2b_{11}\psi_{11} & 0 & -2b_{21}\psi_{21} & 2b_{11}\theta_{11} & 0 & 2b_{21}\theta_{21}\\
2b_{12}\psi_{11} & -a_{11} & B_{22}\psi_{21} & B_{22}\psi_{12} & 2b_{12}\theta_{02}+c_1 & B_{22}\theta_{21} & B_{22}\theta_{12}\\
0 & -B_{21}\psi_{21} & -a_{12} & -B_{21}\psi_{11} & -B_{21}\theta_{21} & c_1 & -B_{21}\theta_{11}\\
2b_{22}\psi_{21} & -B_{12}\psi_{12} & B_{12}\psi_{11} & -a_{21} & B_{12}\theta_{12} & -B_{12}\theta_{11} & 2b_{22}\theta_{02}+c_2\\
0 & -\text{Pr Ray } c_1 & 0 & 0 & \text{Pr } a_{11}^2 & -B_{22}A_{22}\psi_{21} & -B_{22}A_{22}\psi_{12}\\
0 & 0 & -\text{Pr Ray } c_1 & 0 & B_{12}A_{21}\psi_{21} & \text{Pr } a_{12}^2 & B_{21}A_{21}\psi_{11}\\
0 & 0 & 0 & -\text{Pr Ray } c_2 & B_{12}A_{12}\psi_{12} & B_{21}A_{12}\psi_{11} & \text{Pr } a_{21}^2
\end{vmatrix}
\tag{4.115}
$$

where

$$
\begin{aligned}
A_{12} &:= a_{11}-a_{12}, & A_{21} &:= a_{21}-a_{11}, & A_{22} &:= a_{21}-a_{12};\\
B_{12} &:= b_{12}+b_{11}, & B_{21} &:= b_{11}+b_{21}, & B_{22} &:= b_{22}-b_{11}.
\end{aligned}
\tag{4.116}
$$

In the special case of solution (4.111) this simplifies, and only two entries are left in row 5. One of them is eliminated by adding a multiple of the 5th column to the second column. The remaining entry is $a_{11}^2 \neq 0$, defined in (4.77).

After expansion of this determinant with respect to row five we are left with a 6x6 determinant

$$
\begin{vmatrix} A & 0 \\ 0 & B \end{vmatrix},
\tag{4.117}
$$

where A is a 2x2 matrix which is computed to be non-singular for all Ray $\geq R_{cr}^{11}(\alpha)$, i.e. in the domain of existence of solution (4.111). Thus we have to find those parameter values for which B turns singular. Expansion of det(B) is a bit cumbersome, but finally yields the following formula, which was simplified by using expressions for critical curves and ordered by powers of Ray and Pr:

$$
\begin{aligned}
&\frac{B_{12}^2 a_{02} c_1^2}{4 b_{11} b_{12} a_{11}^2} A_{21} \left[\Pr c_1 c_2 - \frac{B_{12}^2 a_{02} c_1^2}{4 b_{11} b_{12} a_{11}^2} A_{12} \right] \text{Ray}^2 + \\
+ & \left[\Pr^2 \left(\frac{B_{12}^2 a_{02} c_1}{4 b_{11} b_{12}} (a_{12}^2 - a_{11}^2)(c_2 + c_1 \frac{a_{21}^2}{a_{11}^2}) + c_1^2 c_2^2 (R_{cr}^{11} - R_{cr}^{21}) \right) + \right. \\
- & \Pr \frac{B_{12}^2 a_{02} c_1}{4 b_{11} b_{12}} (c_1 a_{21} A_{12} + c_2 a_{12} A_{21} - c_1^2 c_2 R_{cr}^{11} \frac{A_{21}}{a_{11}^2}) + \\
- & \left. \frac{B_{12}^2 a_{02} c_1^2}{4 b_{11} b_{12} a_{11}^2} A_{12} A_{21} (a_{12} a_{21} - 2 R_{cr}^{11} \frac{B_{12}^2 a_{02} c_1^2}{4 b_{11} b_{12} a_{11}^2}) \right] \text{Ray} + \qquad (4.118) \\
- & \Pr^2 \left(c_1^2 c_2^2 R_{cr}^{12} (R_{cr}^{11} - R_{cr}^{21}) + R_{cr}^{11} \frac{B_{12}^2 a_{02} c_1}{4 b_{11} b_{12}} (a_{12}^2 - a_{11}^2)(c_2 + c_1 \frac{a_{21}^2}{a_{11}^2}) \right) + \\
+ & \Pr \frac{B_{12}^2 a_{02} c_1}{4 b_{11} b_{12}} R_{cr}^{11} \left(c_1 A_{12} (a_{21} + c_1 c_2 \frac{R_{cr}^{11}}{a_{11}^2}) + c_2 a_{12} A_{21} \right) + \\
+ & R_{cr}^{11} \frac{B_{12}^2 a_{02} c_1^2}{4 b_{11} b_{12} a_{11}^2} A_{21} A_{12} \left(a_{12} a_{21} - R_{cr}^{11} \frac{B_{12}^2 a_{02} c_1^2}{4 b_{11} b_{12} a_{11}^2} \right) = 0.
\end{aligned}
$$

Intermediate steps of these calculations are given in [Mat90, LMP, Pe88].

For each given α and Pr eq. (4.118) gives the Ray value $R_{pr}^{11}(\alpha)$ for which there is a secondary bifurcation on the 2-roll solution. These are the bifurcation points of the mixed-mode solutions described by Busse and Or and discussed in the previous chapter.

In Fig. 4.6 we show these Ray values as functions of α^{-1}, for two different values of Pr. They were obtained through eigenvalue computation of the matrix given in (4.115) and of its 9×9 companion, as explained in the paragraph 'the perturbation approach'. The deviations between the values of the 7×7 model and of the 9×9 model are so small that they cannot be seen in this scale. In [Mat90, LMP, Pe88] the curves $R_{pr}^{11}(\alpha^{-1})$ were plotted by numerical evaluation of other versions of formula (4.118) for several values of Pr with a root-finding algorithm.

For the limiting cases Pr $\to 0$ and Pr $\to \infty$ formula (4.118) simplifies considerably:

Since we introduced R_{cr}^{11} into the formula, it is easy to see that $\text{Ray}(\alpha) = R_{cr}^{11}(\alpha)$ yields a nontrivial solution of eq. (4.118) for $\text{Pr} = 0$ and all appropriate α.

The expression for $\text{Pr} \to \infty$ is obtained by neglecting all terms that do not contain a factor Pr^2. With the abbreviations

$$\gamma_1 := \frac{B_{12}^2 a_{02} c_1}{4 b_{11} b_{12}} (a_{12}^2 - a_{11}^2)(c_2 + c_1 \frac{a_{21}^2}{a_{11}^2}), \quad \delta_1 := c_1^2 c_2^2 (R_{cr}^{11} - R_{cr}^{21}) \quad (4.119)$$

we obtain

$$R_{\infty}^{11}(\alpha) = \frac{\gamma_1(\alpha) + \delta_1(\alpha) R_{cr}^{12}(\alpha) / R_{cr}^{11}(\alpha)}{\gamma_1(\alpha) + \delta_1(\alpha)} R_{cr}^{11}(\alpha). \quad (4.120)$$

If we evaluate eq. (4.120) in $\alpha = \alpha_{21}^{11}$, we find that δ_1 vanishes. We thus can see analytically that $R_{\infty}^{11}(\alpha_{21}^{11}) = R_{cr}^{11}(\alpha_{21}^{11}) = R_{cr}^{21}(\alpha_{21}^{11})$, i.e. that the curve $R_{\infty}^{11}(\alpha)$ emanates from the intersection of R_{cr}^{11} with R_{cr}^{21}. We cannot make any statement, however, about "where this curve goes" for large Ray (note that R_{cr}^{11} and R_{cr}^{12} do not intersect).

It was found that the solution obtained numerically from the formula corresponding to eq. (4.118) for $\text{Pr} = 10^3$ coincides up to plotting accuracy with $R_{\infty}^{11}(\alpha)$, while the one for $\text{Pr} = 10^{-3}$ coincides with $R_{cr}^{11}(\alpha)$ [LMP]. Later on it was found that a writing error was introduced into the calculations of [Pe88, LMP], and it propagated. It falsified the formula corresponding to (4.118) and the curves for $\text{Pr} = 0.5$ and $\text{Pr} = 1.$ in [LMP, Fig. 2], but it did not affect the limiting cases $\text{Pr} = 0$ and $\text{Pr} = \infty$.

4.4.3 Calculation of secondary bifurcation points on the 4-roll solutions

Starting from determinant (4.115), we proceed similarly for the 4-roll solution involving ψ_{21}, θ_{21} and θ_{02}. This time it is the last row which contains only two non-zero entries. By exchanging the last with the 2nd column and then the 2nd row with the 4th row, we get a more convenient version of the matrix. We now add a multiple of the 4th column to the 2nd column and thus eliminate one of the entries in the last row. Expansion with respect to the last row leaves us with a 6 x 6 determinant, which again has the structure shown in (4.117). This time matrix A contains expressions in ψ_{21}, θ_{21} and θ_{02}, and its determinant can be shown to be non-zero for all Ray and α for which solution (4.112) exists. Thus we are left with a 4 x 4 matrix,

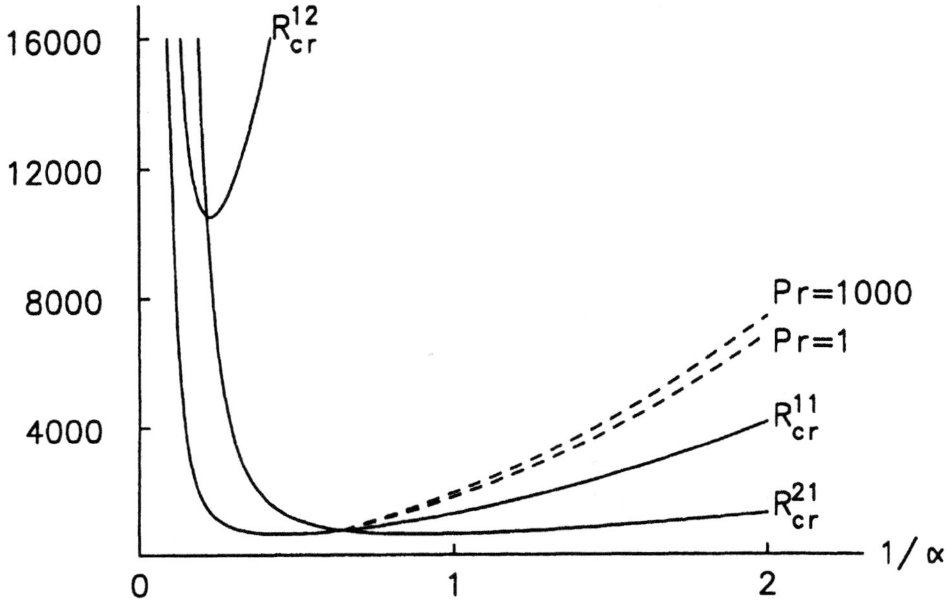

Figure 4.6: Loci of secondary bifurcation points on the 2-roll solution in the $(\alpha^{-1}, \mathrm{Ray})$ plane, for the values Pr = 1 and Pr = 1000; obtained through eigenvalue computation both for the 7×7 and for the 9×9 model.

which is nearly full and thus produces a lengthy expression. Ordering by powers of Ray and Pr and using the expressions for critical curves, we get

$$\frac{B_{21}B_{22}a_{02}c_2^2}{4b_{21}b_{22}a_{21}^2}A_{21}\left[\Pr\, c_1^2+\frac{B_{21}B_{22}a_{02}c_2^2}{4b_{21}b_{22}a_{21}^2}A_{22}\right]\mathrm{Ray}^2$$
$$+\left[\Pr^2\left(\frac{B_{21}B_{22}a_{02}}{4b_{21}b_{22}}\left(c_1c_2(a_{12}^2+a_{11}^2)+c_1^2a_{21}^2+c_2^2\frac{a_{11}^2a_{12}^2}{a_{21}^2}\right)+c_1^4(R_{cr}^{21}-R_{cr}^{11})\right)\right.$$
$$+\Pr\,\frac{B_{21}B_{22}a_{02}c_1c_2}{4b_{21}b_{22}}(a_{11}A_{22}+a_{12}A_{21}+c_1c_2R_{cr}^{21}\frac{A_{22}}{a_{21}^2})$$
$$\left.+\frac{B_{21}B_{22}a_{02}c_2^2}{4b_{21}b_{22}a_{21}^2}A_{22}A_{21}(a_{11}a_{12}-2R_{cr}^{21}\frac{B_{21}B_{22}a_{02}c_2^2}{4b_{21}b_{22}a_{21}^2})\right]\mathrm{Ray}$$
$$-\Pr^2\left(c_1^4R_{cr}^{12}(R_{cr}^{21}-R_{cr}^{11})+R_{cr}^{21}\frac{B_{21}B_{22}a_{02}}{4b_{21}b_{22}}\right.$$
$$\left.\times\left(c_1c_2(a_{12}^2+a_{11}^2)+c_1^2a_{21}^2+c_2^2\frac{a_{11}^2a_{12}^2}{a_{21}^2}\right)\right)$$

$$- \Pr \frac{B_{21}B_{22}a_{02}c_1c_2}{4b_{21}b_{22}} R_{cr}^{21} \left(A_{22}(a_{11} + c_1c_2 \frac{R_{cr}^{21}}{a_{21}^2}) + a_{12}A_{21} \right)$$

$$- R_{cr}^{21} \frac{B_{21}B_{22}a_{02}c_2^2}{4b_{21}b_{22}a_{21}^2} A_{21}A_{22} \left(a_{11}a_{12} - R_{cr}^{21} \frac{B_{21}B_{22}a_{02}c_2^2}{4b_{21}b_{22}a_{21}^2} \right) = 0. \qquad (4.121)$$

Intermediate steps of these calculations are given in [Mat90, LMP, Pe88]. Figure 4.7 shows the curve R_{pr}^{21} versus α^{-1} for the values $\Pr = 1$ and $\Pr = 1000$, obtained through computation of eigenvalues as explained in the paragraph 'the perturbation approach'.

The limiting cases $\Pr \to 0$ and $\Pr \to \infty$ can be obtained analytically again. We find that $\mathrm{Ray}(\alpha) = R_{cr}^{21}(\alpha)$ yields a nontrivial solution of eq. (4.121) for $\Pr = 0$ and all α.

The expression for $\Pr \to \infty$ is obtained again by neglecting all terms that do not contain a factor $\Pr^2$. With the abbreviations

$$\gamma_2 := \frac{B_{21}B_{22}a_{02}}{4b_{21}b_{22}} \left(c_1c_2(a_{12}^2 + a_{11}^2) + c_1^2a_{21}^2 + c_2^2 \frac{a_{11}^2a_{12}^2}{a_{21}^2} \right) \qquad (4.122)$$

$$\delta_2 := c_1^4(R_{cr}^{21} - R_{cr}^{11})$$

we obtain

$$R_{\infty}^{21}(\alpha) = \frac{\gamma_2(\alpha) + \delta_2(\alpha) R_{cr}^{12}(\alpha)/R_{cr}^{21}(\alpha)}{\gamma_2(\alpha) + \delta_2(\alpha)} R_{cr}^{21}(\alpha). \qquad (4.123)$$

$R_{\infty}^{21}(\alpha)$ is the locus of secondary bifurcation points on the 4-roll solution for $\Pr = \infty$. If we evaluate eq. (4.123) in $\alpha = \alpha_{21}^{11}$, we find that δ_2 vanishes. We thus can see analytically that $R_{\infty}^{21}(\alpha_{21}^{11}) = R_{cr}^{21}(\alpha_{21}^{11}) = R_{cr}^{11}(\alpha_{21}^{11})$, i.e. that the curve $R_{\infty}^{21}(\alpha)$ emanates from the intersection of R_{cr}^{11} with R_{cr}^{21}.
If we evaluate eq. (4.123) in $\alpha = \alpha_{12}^{21}$, we find that $R_{\infty}^{21}(\alpha_{12}^{21}) = R_{cr}^{21}(\alpha_{12}^{21}) = R_{cr}^{12}(\alpha_{12}^{21})$, i.e. that the curve $R_{\infty}^{21}(\alpha)$ emanates also from the intersection of R_{cr}^{21} with R_{cr}^{12}. The curve $R_{\infty}^{21}(\alpha)$ thus connects the intersection point of $R_{cr}^{11}(\alpha)$ and $R_{cr}^{21}(\alpha)$ with the intersection point of $R_{cr}^{21}(\alpha)$ and $R_{cr}^{12}(\alpha)$, as was speculated by Meyer-Spasche, Wagner in the case of the Taylor problem (Fig. 3.24).

Figure 4.7 shows the numerically obtained curves $R_{pr}^{21}(\alpha)$ for $\Pr = 1$ and $\Pr = 1000$. It was obtained through eigenvalue computation, as described in the paragraph 'the perturbation approach'. Similar curves were given in [LMP, Fig. 3] and in [Pe88] and [Mat90]. They were obtained from eq. (4.121) by a root-finding algorithm and look the same. ψ_{21} and θ_{21} of (4.112) are real only for $\mathrm{Ray} \geq R_{cr}^{21}$. We thus do not show $R_{\infty}^{21}(\alpha)$ beyond α_{21}^{12} and α_{21}^{11} though the curve given in eq. (4.123) does exist there.

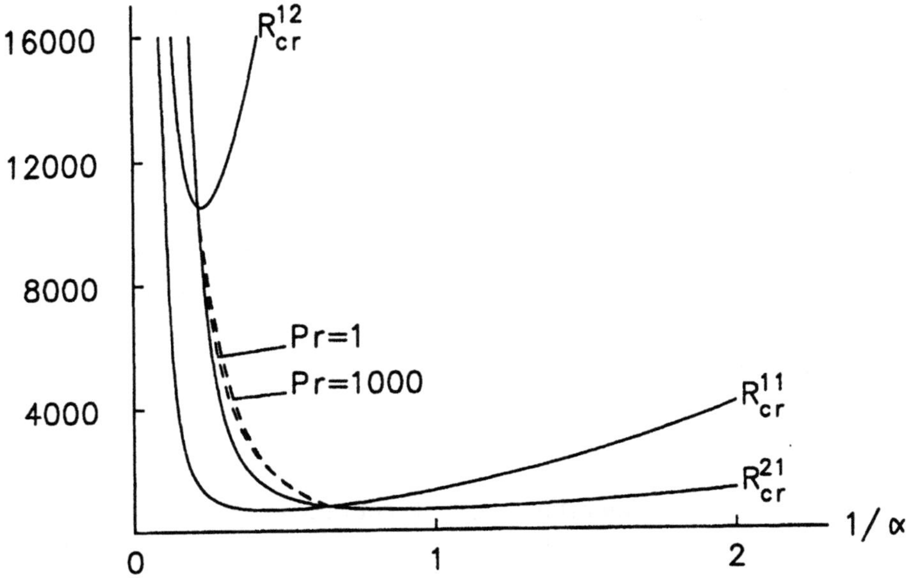

Figure 4.7: Loci of secondary bifurcation points on the 4-roll solution in the $(\alpha^{-1}, \mathrm{Ray})$ plane, for the values Pr = 1 and Pr = 1000; obtained through eigenvalue computation both for the 7×7 and for the 9×9 model.

4.4.4 The perturbation approach

If the full system (4.74)–(4.76) is reduced for local investigations of mode interactions by a perturbation approach to third order in $(\lambda - \lambda_o)^{1/2}$ and $(\mathrm{Ray} - \mathrm{Ray}_o)^{1/2}$, $\lambda = 2\pi/\alpha$, the resulting system involves the 7 modes and equations we used in our model problem (4.114), plus two more: the modes ψ_{32} and θ_{32} and eqs. (4.83) and (4.89). Interactions of the kth and the $(k+1)$st mode were investigated by Knobloch, Guckenheimer [KG83] in a neighborhood of the intersection point $(\alpha_{k1}^{(k+1)1}, \mathrm{Ray}_{k1}^{(k+1)1})$. In the case $k = 1$, the system of equations they used [KG83, (15), (16)] is identical with the system we get from our 7×7 system by including ψ_{32} and θ_{32}. Note, however, that the coefficients of the system vary with α in our approach, but not with time t.

Knobloch/Guckenheimer did not calculate loci of secondary bifurcation points. Indeed, we doubt that the 9×9 system could be treated the way the 7×7 system was treated: not even when using such a device as Mathematica. Both calculations of us [LMP, Mat90] actually do contain minor errors which are eliminated in this text. In a recent investigation (joint work with F. Pohl), we used the methods to be discussed next to reproduce the results on the 7×7 model problem before we checked the question how much the results change when the two additional modes and equations are included.

The method employed avoids the computation of determinants: the eigenvalues both of the 7×7 matrix and of the 9×9 matrix were computed with standard software (RG of Eispack), and the critical values were found essentially by bisection or Newton iterations. The curves of critical points were then obtained by linear interpolation between the computed points. The results are shown in Figs. 4.6 and 4.7.

We found that

- both previous calculations do not contain errors that would have changed the results qualitatively, and there are no visible quantitative deviations from the results of [Mat90];
- each two curves corresponding to the 7×7 model and to the 9×9 model, respectively, do differ, but below plotting accuracy in Figs. 4.6 and 4.7.

For several points in the parameter plane the eigenvalues of corresponding matrices were compared in detail.

Critical values on the 2-roll solution:

The largest relative deviations between the values for 7×7 and 9×9 were found for $\alpha^{-1} = 2$, the smallest for $\alpha \approx \alpha_{11}^{21}$. For $\alpha^{-1} = 2$ we obtained the following values for the critical Rayleigh number Ray_c:

$$\begin{array}{llllll} \mathrm{Pr} = 1: & 7 \times 7 & 9 \times 9 & \qquad \mathrm{Pr} = 1000: & 7 \times 7 & 9 \times 9 \\ & 6901.92 & 6842.63 & & 7511.86 & 7420.83. \end{array} \tag{4.124}$$

When normalized with the Ray_c-value of the 9×9 problem for that α, we found deviations less than $9 \cdot 10^{-3}$ for $\mathrm{Pr} = 1$ and less than $1.2 \cdot 10^{-2}$ for $\mathrm{Pr} = 1000$ for $\alpha^{-1} = 2$, and deviations less than 10^{-5} for both values of Pr and $\alpha = 0.645$. Figure 4.8 shows an enlarged portion of Fig. 4.6, produced using 5 α-values and linear interpolation. For $\alpha^{-1} < 1.0$, the 7×7- and 9×9-values still coincide when plotted in the scale of Fig. 4.8. The eigenvalues

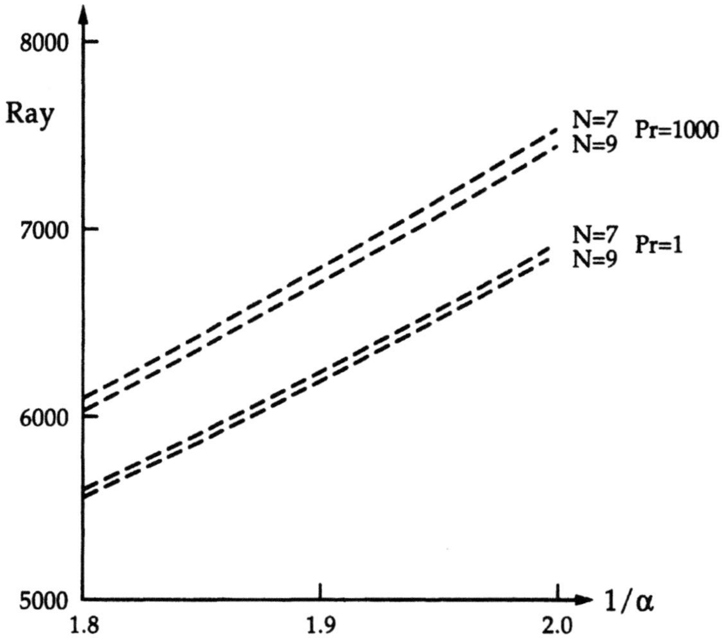

Figure 4.8: Loci of secondary bifurcation points on the 2-roll solution in the $(\alpha^{-1}, \mathrm{Ray})$ plane as obtained through eigenvalue computation of the 7×7 and 9×9 models, with linear interpolation between 5 α-values. Enlarged portion of Fig. 4.6.

in the (α, Ray)-values of (4.124) were computed to be:

Pr =1:		Pr =1000:		
5.797'-9	-7.542'-9	4.910'-10	-7.508'-9	
-1.919'+1	-2.018'+1	-1.433'+1	-1.457'+1	
-8.497'+0	-8.371'+0	-9.403'+0	-9.224'+0	
2.242'+1	2.226'+1	2.197'+1	2.179'+1	
9.094'+1	9.097'+1	1.024'+5	1.024'+5	
9.275'+1	9.495'+1	1.181'+5	1.181'+5	
1.572'+3	1.572'+3	1.578'+6	1.578'+6	
–	-5.106'+1	–	-5.109'+1	
–	1.749'+3	–	1.741'+6	(4.125)

All of them are different for all 4 matrices, but it is clear how they correlate and which one would change into which in a parameter continuation process.

In the case of $\alpha^{-1} = 0.645$ and $\mathrm{Pr} = 1000$, one eigenvalue of the 7×7 matrix and the 9×9 matrix is below 10^{-8} for Ray values with relative deviation below 10^{-5}. Another one is of size $2.4 \cdot 10^{-2}$, to become zero in the intersection point. The two approximations to it have a relative deviation below $2 \cdot 10^{-2}$, all other relative deviations are below 10^{-5}. For $\mathrm{Pr} = 1$, the deviations are much smaller.

Critical values on the 4-roll solution:

The findings for the critical values on the 4-roll solution are essentially the same. The eigenvalues are different for the 7×7 and the 9×9 cases, but the relative deviations are small and do not introduce qualitative changes. The largest relative deviations were found near $\alpha^{-1} = 0.339$, the smallest for $\alpha \approx \alpha_{11}^{21}$ and $\alpha \approx \alpha_{12}^{21}$. It had been speculated that the Galerkin truncation would change the curvature of the center manifold at each double point and thus the inclination of the curve of secondary bifurcation points at those double points (Knobloch, private communication, 1991). The numerical model did not show this.

This time we fixed Ray_c and searched for the corresponding α^{-1} value. For $\mathrm{Ray}_c = 3605.11$ we obtained the following values for the critical α_c^{-1}:

$\mathrm{Pr} = 1:$	7×7	9×9	$\mathrm{Pr} = 1000:$	7×7	9×9	(4.126)
	.33926	.33900		.35323	.35269	

The eigenvalues in the (α, Ray)-values of (4.126) were computed to be:

Pr =1:		Pr =1000:		
-1.983'-11	-4.082'-12	-4.422'-10	-4.505'-8	
-1.625'+1	-1.619'+1	-1.898'+1	-1.887'+1	
1.178'+1	1.191'+1	5.042'+0	4.934'+0	
2.894'+1	2.888'+1	3.154'+1	3.144'+1	
2.707'+2	2.685'+2	3.198'+5	3.207'+5	
1.947'+3	1.951'+3	1.758'+6	1.776'+6	
2.315'+3	2.316'+3	2.256'+6	2.258'+6	
–	-1.384'+2	–	-1.324'+2	
–	1.390'+4	–	1.251'+7	(4.127)

All of them are different for all 4 matrices, but it is clear how they correlate and which one would change into which in a parameter continuation process.

4.5 The (2,6) interaction in a model problem

To investigate the interaction of the 2- and 6-roll solutions, there are now the same two ways as in the $(2,4)$-case for obtaining a simplified model problem:

a) make a perturbation approach as just discussed for the $(2,4)$-case. A 3rd order perturbation analysis of the interaction of the (k,1)-mode with the (m,1)-mode shows that the modes (k,1), (m,1), (k+m,2), (m-k,2), (0,2) should be taken into account [Seg62, KG83]. For $k = 1,\ m = 3$ this leads to a model with the modes

$$\psi_{11}, \psi_{22}, \psi_{31}, \theta_{02}, \theta_{11}, \theta_{22}, \theta_{31}, \psi_{42}, \theta_{42}.$$

b) truncate system (4.74)–(4.76) at $N = 2,\quad M = 3$ and use perturbation arguments to further simplify this system. This approach was used by us in [Mat90]. It leads to the following model:

$$\begin{aligned}
(b_{21} - b_{32})(a_{31} - a_{22})\psi_{22}\psi_{31} - \text{Pr Ray } c_1\theta_{11} + \text{Pr } a_{11}^2\psi_{11} &= 0,\\
(b_{12} + b_{21})(a_{11} - a_{22})\psi_{11}\psi_{22} - \text{Pr Ray } c_3\theta_{31} + \text{Pr } a_{31}^2\psi_{31} &= 0,\\
(b_{11} + b_{31})(a_{31} - a_{11})\psi_{11}\psi_{31} - \text{Pr Ray } c_2\theta_{22} + \text{Pr } a_{22}^2\psi_{22} &= 0,\\
-2b_{12}\psi_{11}\theta_{02} + (b_{21} - b_{32})(\psi_{31}\theta_{22} + \psi_{22}\theta_{31}) - c_1\psi_{11} + a_{11}\theta_{11} &= 0,\\
-2b_{32}\psi_{31}\theta_{02} + (b_{12} + b_{21})(\psi_{22}\theta_{11} - \psi_{11}\theta_{22}) - c_3\psi_{31} + a_{31}\theta_{31} &= 0,\\
(b_{11} + b_{31})(\psi_{31}\theta_{11} + \psi_{11}\theta_{31}) - c_2\psi_{22} + a_{22}\theta_{22} &= 0,\\
2b_{11}\psi_{11}\theta_{11} + 2b_{31}\psi_{31}\theta_{31} + a_{02}\theta_{02} &= 0.
\end{aligned} \tag{4.128}$$

The situation is very similar as in the $(2,4)$-case, and we expect that the more accurate results obtained with the 9×9 model will differ only slightly from the results obtained with the 7×7 model, which we will discuss next.

4.5.1 Calculation of secondary bifurcation points on the 2-roll solutions

A necessary condition for secondary bifurcation on a branch of solutions of (4.128) is that the determinant of the Jacobian vanishes. As in the $2,4$ interaction, this determinant simplifies considerably on both solution branches under consideration. In the case of the 2-roll solutions row 5 contains only 2 entries. We divide them by Pr and eliminate one of them. After one expansion of the determinant we are left with a 6×6 determinant of the form (4.117). The 2×2 matrix A is computed to be nonsingular for $\text{Ray} - R_{cr}^{11} \geq 0$,

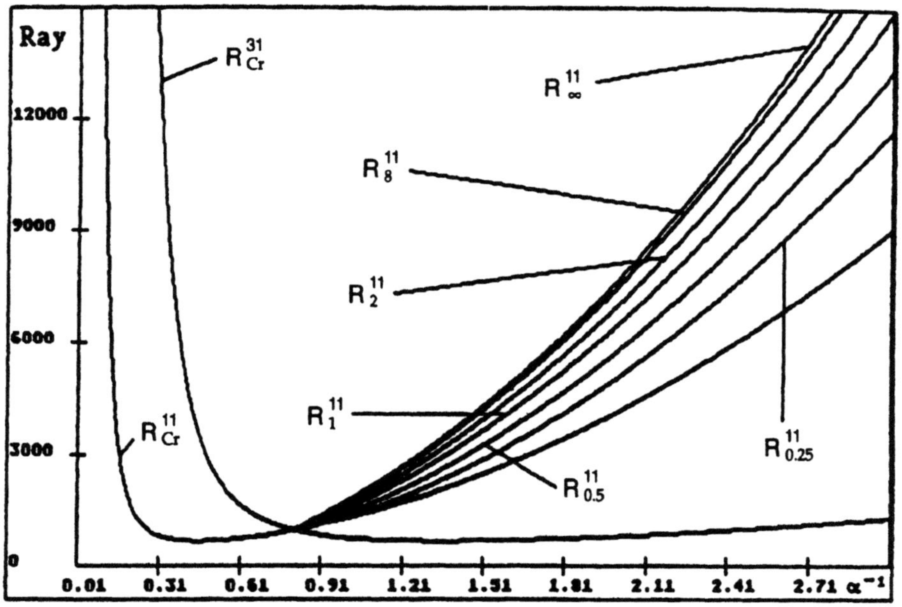

Figure 4.9: Loci of secondary bifurcation points induced on the 2-roll solution by the $(2,6)$ interaction for the indicated values of Pr (after [Mat90, Fig.12a]).

i.e. in the domain of existence of the 2-roll solutions (4.111). We are thus left with a 4×4 determinant, which we actually have to compute. The result is given in [Mat90, (12.7)]. It is a quadratic polynomial in Ray and in Pr and a rational function in α, with a structure very similar to (4.118). For each given pair (α, Pr) this formula gives the Ray-value for which there is a secondary bifurcation on the 2-roll solution.

In Fig. 4.9 these Ray-values are plotted versus α^{-1}, for several Pr values. The curves for $\mathrm{Pr} < \infty$ were obtained numerically with a root-finding algorithm from the formula corresponding to (4.118). Again, there are simple expressions for the limiting cases: for $\mathrm{Pr} = 0$, the curve coincides with R^{11}_{cr}, and for $\mathrm{Pr} \to \infty$ we get

$$R^{11}_{31}(\alpha; \mathrm{Pr} \to \infty) = \frac{\gamma_3 + \delta_3 R^{22}_{cr}/R^{11}_{cr}}{\gamma_3 + \delta_3} R^{11}_{cr} \tag{4.129}$$

with

$$\begin{aligned} \gamma_3 &:= \frac{(b_{12}+b_{21})(b_{11}+b_{31})a_{02}}{4b_{11}b_{12}}(c_1 a_{22}^2 - c_2 a_{11}^2)(c_3 + c_1 \frac{a_{31}^2}{a_{11}^2}) \\ \delta_3 &:= c_2^2 c_3^2 (R^{11}_{cr} - R^{31}_{cr}). \end{aligned} \tag{4.130}$$

Note that δ_3 vanishes at $\alpha = \alpha_{31}^{11}$, i.e. the curve $R_{31}^{11}(\alpha; \mathrm{Pr} \to \infty)$ emanates from the double point $(\alpha_{31}^{11}, \mathrm{Ray}_{31}^{11})$, as was to be expected from the previous section and from Chapter 3. R_{cr}^{22} plays a prominent role in formula (4.129), but $R_{cr}^{22}(\alpha)$ and $R_{cr}^{11}(\alpha)$ do not intersect. As in the case of the $(2,4)$ interaction, we cannot say anything about 'where this curve goes'. Maybe it continues to infinity, maybe it ends in some other singular point which has been neglected in our simplified model.

4.5.2 Calculation of secondary bifurcation points on the 6-roll solutions

To find the secondary bifurcation points on the 6-roll solutions, we start from the determinant of the Jacobian of system (4.128) again. Quite a number of terms turn into zero when solution (4.113), $p = 3$, is inserted. The resulting pattern of zeros is inconvenient, but very similar to the pattern in the 4-roll case. Interchanges of columns and rows, elimination of an entry and one expansion produce again a 6×6 determinant of form (4.117), and the 2×2 matrix A can be shown to be nonsingular in the (α, Ray) domain of existence of the 6-roll solutions. Thus we have to compute a 4×4 determinant again, and the result is given in [Mat90, (13.8)]. Its structure is very similar to the one of (4.121) in the $(2,4)$ case. The curves for $\mathrm{Pr} = 1.$ and for $\mathrm{Pr} \to \infty$ are shown in Fig. 4.10. Again, the curve for $\mathrm{Pr} = 1$ had to be obtained numerically, while there are simple expressions for the limiting cases: for $\mathrm{Pr} = 0$ the curve coincides with R_{cr}^{31}, and for $\mathrm{Pr} \to \infty$ it becomes

$$R_{11}^{31}(\alpha; \mathrm{Pr} \to \infty) = \frac{\gamma_4 + \delta_4 R_{cr}^{22}/R_{cr}^{31}}{\gamma_4 + \delta_4} R_{cr}^{31} \tag{4.131}$$

with

$$\gamma_4 := \frac{(b_{32} - b_{21})(b_{11} + b_{31})a_{02}}{4b_{31}b_{32}} \left(c_2 c_3 a_{11}^2 + c_1 c_3 a_{22}^2 + c_1 c_2 a_{31}^2 + c_3^2 \frac{a_{22}^2 a_{11}^2}{a_{31}^2} \right)$$

$$\delta_4 := c_1^2 c_2^2 (R_{cr}^{31} - R_{cr}^{11}).$$

The curve given in (4.131) connects the two double points $(\alpha_{31}^{11}, \mathrm{Ray}_{31}^{11})$ and $(\alpha_{31}^{22}, \mathrm{Ray}_{31}^{22})$.

Computations and plotting by Matern were done on an ATARI-1040ST (rounding error level at 10^{-5}). Matern's results on the (2,6) interaction remained unpublished until now because their value was doubtful without

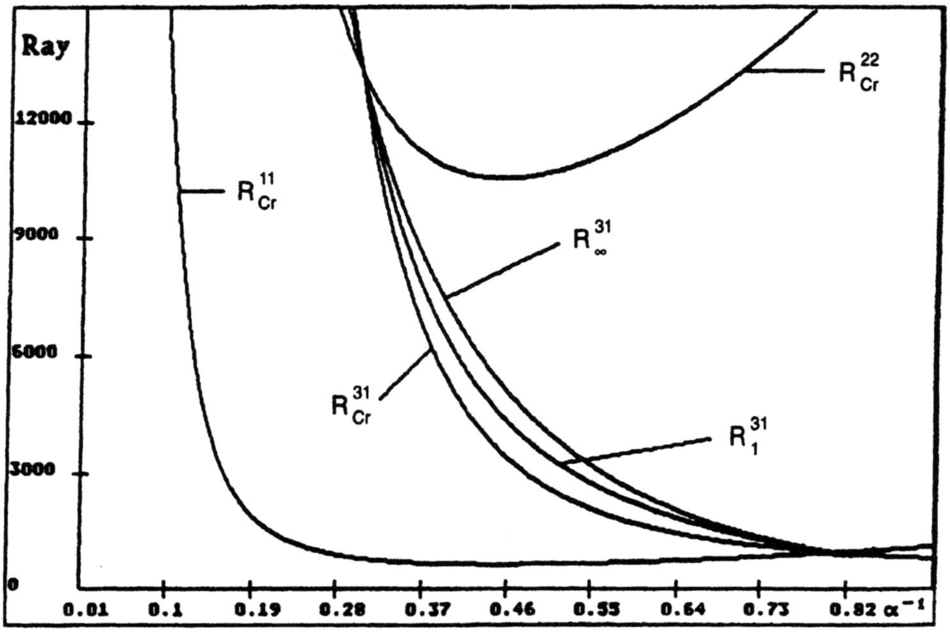

Figure 4.10: Loci of secondary bifurcation points on the 6-roll solution in the $(\alpha^{-1}, \text{Ray})$ plane for $\text{Pr} = 1$ and $\text{Pr} = \infty$ (after [Mat90, Fig.13a]).

the comparison of the 7×7 and 9×9 model problems in the previous paragraph.

4.5.3 Nonlinear interactions between the bifurcating branches

From some of the Taylor vortex computations discussed in Chapter 3 it was expected that there are nonlinear interactions between the $(2, 4)$ and the $(2, 6)$ mixed-mode solutions. In computations with fixed wavelengths and varying Reynolds number, two different solution branches clearly could be identified to be generated by the (2,4)- and by the (2,6)-double point, respectively. Then holding the Reynolds number fixed and varying the wavelength in another set of computations, solutions on one of the branches could be changed smoothly into solutions on the other branch. This was explained by a perturbed bifurcation between the two branches, but it was unclear where to locate the center of the interaction (see Figs. 3.28, 3.29, 3.31 and their discussion).

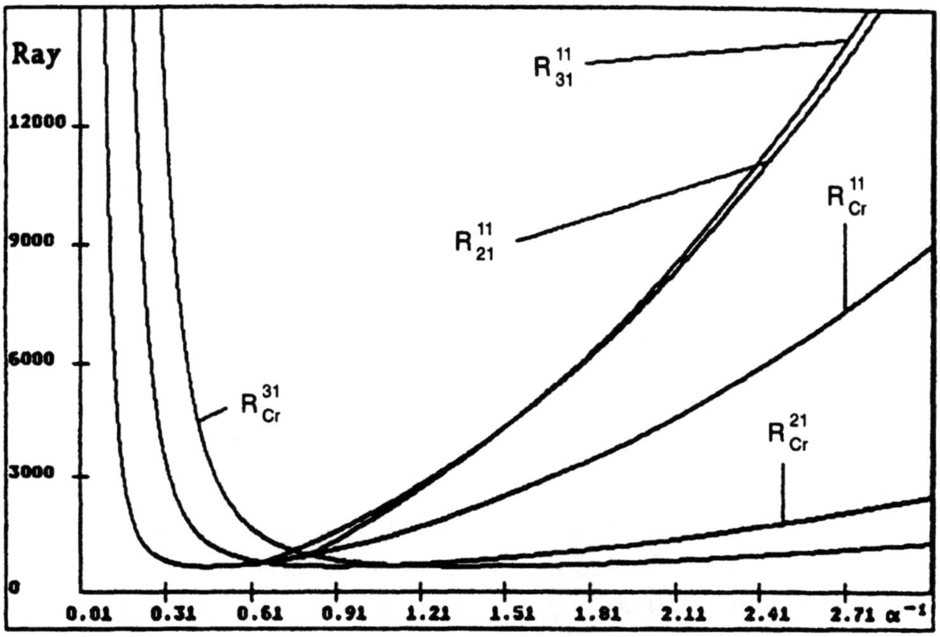

Figure 4.11: Loci of secondary bifurcation points on the 2-roll solution in the $(\alpha^{-1}, \mathrm{Ray})$ plane. Shown are the curves induced by the $(2,4)$ and by the $(2,6)$ interaction, for infinite Pr (after [Mat90, Fig.15a]).

To locate an organizing center of these interactions, it does not make sense to look at the curves (4.131) and (4.123): these curves lie on two different solution surfaces. The two curves (4.129) and (4.120), however, both lie on the surface of 2-roll solutions, and they do intersect. The intersection point is shown in Fig. 4.11, which was obtained by merging parts of Fig. 4.9 with Matern's version of Fig. 4.6. For $\mathrm{Pr} < \infty$ the curves intersect in a very similar way, but the inclination of both curves differs less for smaller Pr [Mat90, Fig. 15b ($\mathrm{Pr} = 8$); Fig. 15c ($\mathrm{Pr} = 10$)]. In the limiting case $\mathrm{Pr} = 0$ both curves coincide with each other, because they coincide with R_{cr}^{11}, as we have seen earlier.

Our computations of Taylor vortex flows made us speculate that such intersection points exist and that curves of tertiary bifurcations eminate from them, see the text near Figs. 3.29 and 3.33.

4.6 Generalisations and consequences

4.6.1 Other interactions

As we have seen, the modes $(k, 1)$, $(m, 1)$, $(k+m, 2)$, $(m-k, 2)$, $(0, 2)$ have to be taken into account in a perturbation approach to 3rd order to investigate the interaction of the $(k, 1)$-mode with the $(m, 1)$-mode. Comparison of models a) and c) approximating the Rayleigh-Bénard problem showed that modes $(k+m, 2)$ did not change results qualitatively. For finding the basic pattern, it thus seems to be sufficient to investigate systems of 7 equations composed by the modes $(k, 1)$, $(m, 1)$, $(m-k, 2)$, $(0, 2)$. The structures of the resulting determinants were found to be very similar for the 2-roll, 4-roll and the 2-roll, 6-roll cases, because the underlying equations (4.74)–(4.76) have the same structure with respect to the different modes.

So we think that the general case $(1, 1)$ with $(m, 1)$ will follow the pattern found here. We expect that there are two curves emenating from each of the double points $(\alpha_{m1}^{11}, \mathrm{Ray}_{m1}^{11})$, $m > 1$. One of the curves connects $(\alpha_{m1}^{11}, \mathrm{Ray}_{m1}^{11})$ with $(\alpha_{m1}^{m-1,2}, \mathrm{Ray}_{m1}^{m-1,2})$, the other curve either goes to infinity, or interacts with a double point excluded by our Galerkin approximation and also by the perturbation approach to third order.

Other interactions can be investigated in a very similar way if it turns out that the computed curves of secondary bifurcations stay close to the critical curves, as they do in the cases investigated here: remember that they merge with the critical curves for $\mathrm{Pr} \to 0$. So we expect that there will be two curves emenating from $(\alpha_{m1}^{k1}, \mathrm{Ray}_{m1}^{k1})$, $m > k$, and that they will be found to go to the double point $(\alpha_{m1}^{m-k,2}, \mathrm{Ray}_{m1}^{m-k,2})$ and to infinity, respectively. Especially we expect that the two curves emenating from $(\alpha_{31}^{21}, \mathrm{Ray}_{31}^{21})$ (Point I in Fig. 3.30) will be found to go to the double point $(\alpha_{31}^{12}, \mathrm{Ray}_{31}^{12})$ and to infinity, respectively. As we can see from Fig. 4.4, this will have the consequence that one of them will meet the curve of secondary bifurcations connecting $(\alpha_{31}^{11}, \mathrm{Ray}_{31}^{11})$ with $(\alpha_{31}^{22}, \mathrm{Ray}_{31}^{22})$, and the two will probably interact.

But the question if neglecting the $(k+m, 2)$-modes is appropriate has to be newly investigated. It should be kept in mind that this strict truncation to systems of 7 or 9 equations was done to find basic patterns. There probably are higher order interactions which got excluded in the simplified models studied here.

Even when restricting the analysis to these basic patterns, coalescence of three and more points of secondary bifurcation has to be expected, and this will give rise to new patterns of solutions. A glimpse into the zoo to be expected is given by some of the figures in Chapter 3, and more by figures displayed in [WS87]. Many of those were not published because it was unclear which interactions generate them.

In the next subsection we show that the superposition of only two or three modes is enough to produce free stagnation points, separatrices and triangular vortices.

4.6.2 Linear superpositions of pure-mode solutions

As we have seen in Chapter 3, many different interactions play a role. So it might be surprising at last that many of the patterns displayed in the figures of Chapter 3 can be produced *essentially* by linear superposition of only three Fourier modes, as is suggested by the analysis of the present chapter (neglect the distorsions caused by the broken symmetry (4.68) when looking at the figures). We give two examples here:

$$(1-s)\ ((1-t)\ f_{11} + t\ f_{21}) - s\ f_{12} \tag{4.132}$$

is displayed in Fig. 4.12, and

$$(1-s)\ ((1-t)\ f_{11} + t\ f_{31}) + s\ f_{22} \tag{4.133}$$

is displayed in Fig. 4.13, where

$$f_{pq}(x,z) := \sin\frac{q\pi}{a}x \quad \sin\frac{2p\pi}{b}z \tag{4.134}$$

are pure-mode solutions defined for $0 \le x < a$ and $-b/2 \le z < b/2$ and $s,\ t = 0(0.25)1$. The superposition factor t changes from left to right, s from top to bottom. While those flows with small s can be observed for small Reynolds numbers, those with larger s can be observed for larger Reynolds number, i.e. increasing the Reynolds number increases the influence of the flow multiplied by s in the mixed-mode solutions. Flows for $t = 0$ with increasing s are not expected to exist according to our analysis since the curves R_{cr}^{11} and R_{cr}^{12} do not intersect and thus do not interact directly. These flows are included here for completeness – such flows were actually never observed.

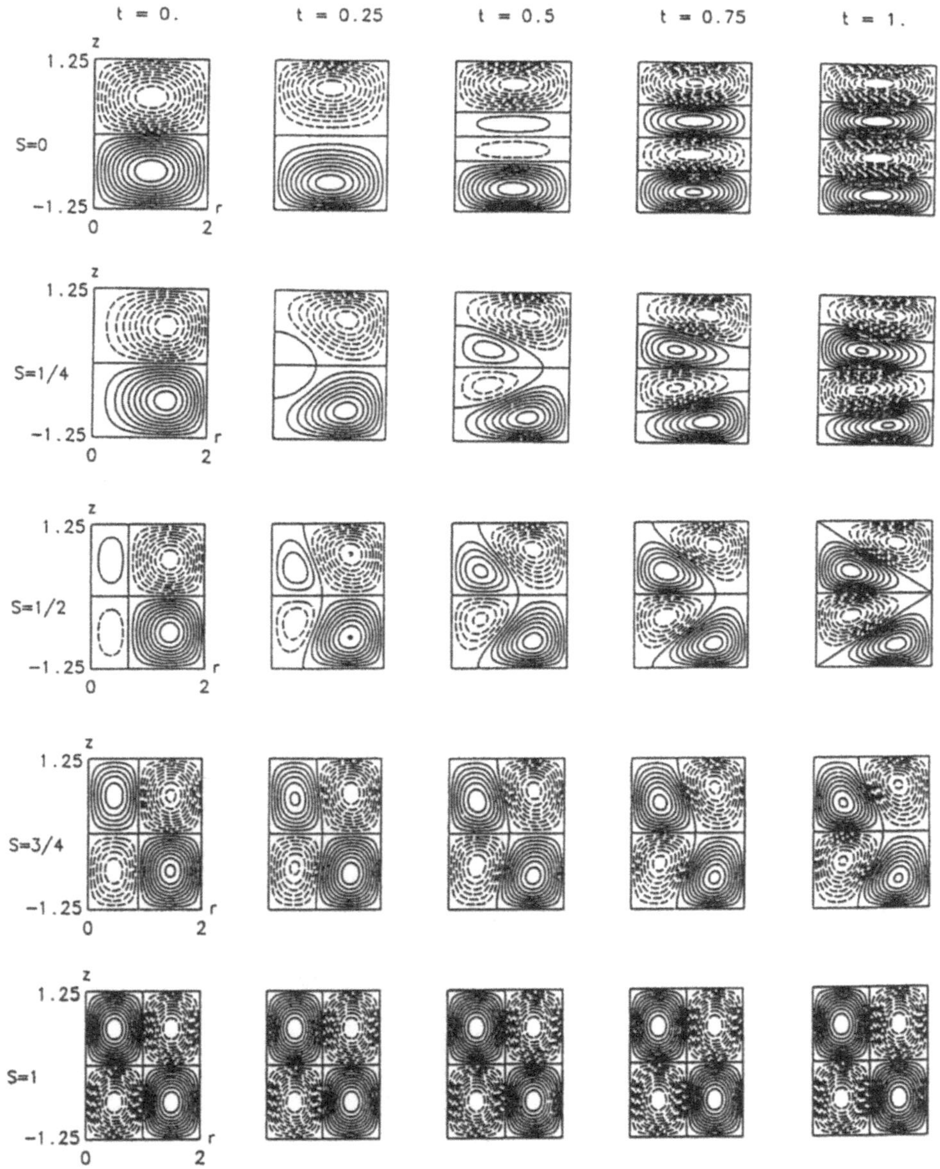

Figure 4.12: Superposition of the three Fourier modes for ψ found to be dominating in the analysis of the 2-roll, 4-roll interaction.
Shown is $g(x, z; s, t) = (1-s)\ ((1-t)\ f_{11} + t\ f_{21}) - s\ f_{12}$. The functions $f_{pq}(x, z)$ are defined in (4.134), and $s,\ t$ vary in steps of 0.25 between 0 and 1 (t from left to right, s from top to bottom).

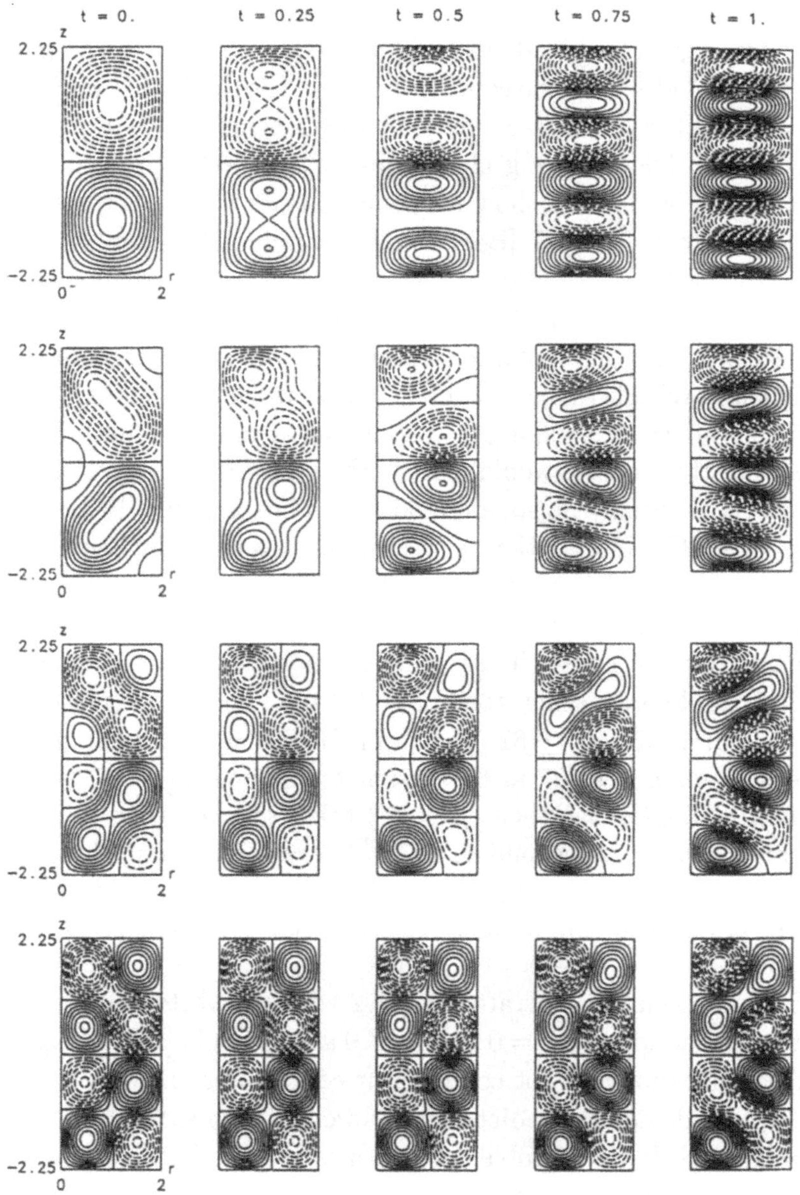

Figure 4.13: Superposition of the three Fourier modes for ψ found to be dominating in the analysis of the 2-roll, 6-roll interaction.
Shown is $g(x,z;s,t) = (1-s)\ ((1-t)\ f_{11} + t\ f_{31}) + s\ f_{22}$. The functions $f_{pq}(x,z)$ are defined in (4.134), and $s,\ t = 0\ (0.25)\ 0.75$ (the rather dull case $s = 1$ with 8 equally sized vortices is omitted).

4.6.3 Secondary bifurcations in the Taylor problem revisited

These investigations should be complimented by a short report on computations by Bolstad. Using the code described in [BoKe, Bol92, Bol92a, Bol93], he computed the loci of secondary bifurcations and folds with bifurcation following and fold-following algorithms in the Taylor problem for several η-values. In Fig. 4.14 we show his computations for the 2-vortex, 4-vortex interaction for $\eta = 0.727$ [Bol]. Similar computations for $\eta = 0.615$ are shown in [Bol92, Fig. 2] ($\mathrm{Re}^\delta \leq 340$) and [Bol92a, Fig. 2] ($\mathrm{Re}^\delta \leq 1100$).

In Chapter 3 we saw that the bifurcations marked (4) and (5) in Figs. 3.4, 3.14 and 3.24 correspond to $\lambda/2$-shifted twins of one secondary bifurcation in the Rayleigh-Bénard problem and that they got shifted to different parameter values by breaking of the midline symmetry (4.68). We thus expected that these curves would meet at the point $(\mathrm{Re}_{d4}, \lambda_{d4})$. In Fig. 4.14 we see that they actually do so, as smoothly as in the point $(\mathrm{Re}_{24}, \lambda_{24})$ shown in Fig. 3.14. The intersection of the curves is due to the projection [Bol]. Unexpected is the detour of curve (4) to the regime of triple-vortex flows, i.e. of flows with 3 vortices in the radial direction. We do not know which other double point causes it. Also, we did not investigate and thus cannot explain the fold emenating from curve (4) near its relative minimum.

The curves for $\eta = 0.615$ shown in [Bol92a, Fig. 2] look quite similar to the ones for $\eta = 0.727$ in the parameter regime up to Re-values corresponding to Ray_{21}^{12}. But then curve (4) takes several other detours, meets the double points corresponding to $(\alpha_{21}^{13}, \mathrm{Ray}_{21}^{13})$ and $(\alpha_{21}^{14}, \mathrm{Ray}_{21}^{14})$, respectively, and disappears to higher Re-values. This was found to be due to a bifurcation between (4) and another secondary curve for $\eta \approx 0.707$, $\mu = 0$ [Bol].

Secondary curves generated by the 2-vortex, 6-vortex and the 4-vortex, 6-vortex interactions for $\eta = 0.615$, $\mu = 0$ are shown in [Bol93, Figs. 1, 3, 4]. These computations do not confirm our conjectures on these interactions. But they also do not contradict them: as we have just seen, the perturbation of the Rayleigh-Bénard problem is too massive for these parameter values to draw any conclusions for the Rayleigh-Bénard problem.

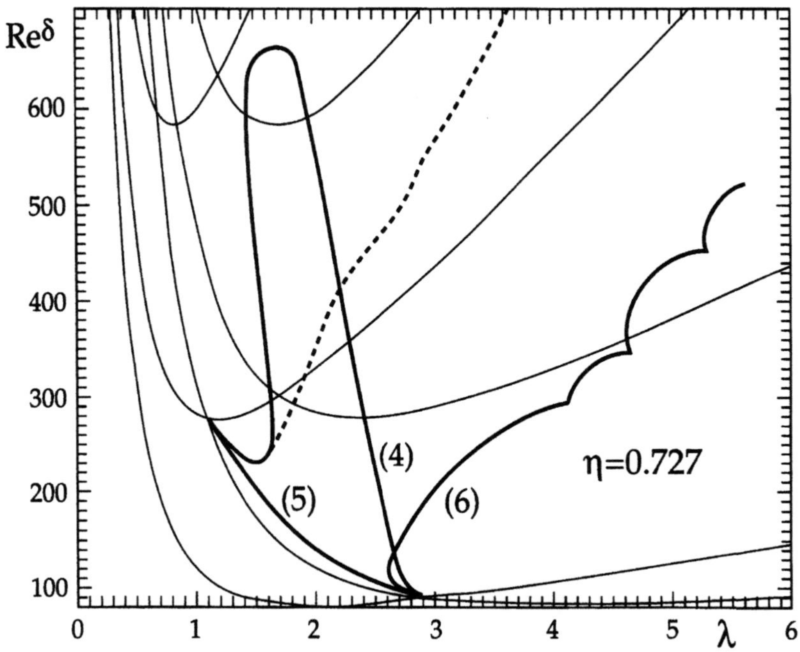

Figure 4.14: The 2-vortex, 4-vortex interaction for $\eta = 0.727$, $\mu = 0$: paths of bifurcation points and of fold points in the $(\lambda, \mathrm{Re}^\delta)$-plane. (4), (5), (6) have the same meaning as in Figs. 3.4, 3.14 and 3.24. Computations by J. Bolstad [Bol].

Bibliography

[AC] G. Ahlers, D.S. Cannell (1984): Wave number selection in rotating Couette-Taylor flow. in [Fre], p. 27–32

[ACDH] G. Ahlers, D.S. Cannell, M.A. Dominguez-Lerma, R. Heinrichs (1986): Wavenumber selection and Eckhaus instability in Couette-Taylor flow. Physica **23** D, 202–219

[AG78] Alziary de Roquefort, T.; Grillaud, G. (1978): Computation of Taylor vortex flow by a transient implicit method Comput. and Fluids **6**, pp.: 259–69

[AH92] *Ordered and Turbulent Patterns in Taylor-Couette Flow*; Proc. NATO Advanced Research Workshop, Columbus, OH, May 22–24, 1991; C.David Andereck, Fernand Hayot, eds.; Plenum Press, New York 1992

[ALS] Andereck, C.D.; Liu, S.S.; Swinney, H.L. (1986): Flow regimes in a circular Couette system with independently rotating cylinders. J. Fluid Mech. **164**, 155–183

[An77] I.P. Andreichikov (1977): Branching of secondary modes in the flow between rotating cylinders. Fluid dynamics **12** (1979), p. 38; translated from Izv. Akad. Nauk SSSR, Mekh.Zhidk. Gaza, No. 1 (1977), 47–53

[An75] I.P. Andreichikov (1975): Calculation of secondary flow between rotating cylinders. Fluid dynamics, translated from Izv. Akad. Nauk SSSR, Mekh.Zhidk. Gaza, No. 2 (1975)

[BKR] L. Bauer, H.B. Keller, E.L. Reiss: Multiple eigenvalues lead to secondary bifurcations. SIAM Reviews 17 (1975), 101–122

[Ben76] T.B. Benjamin (1976): Application of Leray-Schauder degree theory to problems of hydrodynamic stability. Math Proc Camb Phil Soc **79**, 373–392

[Ben78] T.B. Benjamin (1978): Bifurcation phenomena in steady flows of a viscous fluid. I Theory; II Experiments. Proc. Roy. Soc. Lond. A **359**, 1–26; 27–43

[BM] Benjamin, T. B.; Mullin, T. (1982): Notes on the multiplicity of flows in the Taylor experiment. J. Fluid Mech. **121**, 219–230

[BBU87] M.J. Bennett, R.A. Brown, L.H. Ungar (1987) Nonlinear interactions of interface structures of differing wavelength in directional solidification. In: *The Physics of Structure Formation*, W. Güttinger, G. Dangelmayr, eds., Springer Series in Synergetics, (1987), 180–190

[BH79] Blennerhassett, P.J.; Hall, P. (1979): Centrifugal instabilities of circumferential flows in finite cylinders: linear theory. Proc. Roy. Soc. Lond. A **365**, 191–207

[Bol92] J.H. Bolstad (1992): Following paths of symmetry-breaking bifurcation points. LLNL Tech. Rep. UCRL-99530, 1988; Int. J. Bif. Chaos **3** (1992) 559–576

[Bol92a] J.H. Bolstad (1992): Following paths of pitchfork bifurcation points. LLNL Tech. Rep. UCRL-JC-110570, 1992

[Bol93] J.H. Bolstad (1993): Following paths of transcritical bifurcation points. LLNL Tech. Rep. UCRL-JC-113312, March, 1993

[Bol] J.H. Bolstad, private communications 1991, 1993, 1996

[BoKe] J.H. Bolstad, H.B. Keller (1987): Computation of anomalous modes in the Taylor experiment. J. Comp Phys **69**, 230–251

[Boo] O. Booz: Numerische Lösung der Navier-Stokes-Gleichungen für Taylor-Wirbelströmungen im weiten Spalt. Ph.D. dissertation, U. Stuttgart 1980, see also [FBo].

[Bue84] K. Bühler (1984): Der Einfluß einer Grundströmung auf das Einsetzen thermischer Instabilitäten in horizontalen Fluidschichten und die Analogie zum Taylorproblem. Strömungsmechanik und Strömungsmaschinen **34**, 67–76

[BC86] K. Bühler, J.E.R. Coney, M. Wimmer, J. Zierep (1986): Advances in Taylor vortex flow: A report on the fourth Taylor vortex flow working party meeting, Acta Mechanica **62**, 47–61

[Bu72] J.E Burkhalter: Experimental investigation of supercritical Taylor vortex flow. Ph.D. dissertation, U. Texas, 1972, 125 pages

[BuKo] J.E. Burkhalter, E.L. Koschmieder: Steady supercritical Taylor vortices after sudden starts. Phys. Fluids 17(1974), 1929–1935

[Bus83] F.H. Busse (1983): Recent developments in the dynamo theory of planetary magnetism. Ann. Rev. Earth Planet. Sci **11**, 241–286

[Bus85] F.H. Busse (1985): Transition to turbulence in Rayleigh-Bénard convection. [SG, 97-137]

[Bus87] F.H. Busse (1987): Transition to asymmetric convection rolls. in [KST, 18 – 26]

[Bus89] F.H. Busse (1989): Fundamentals of thermal convection. Chapter 2 pp. 23–95 in: *Mantle Convection, Plate Tectonics and Global Dynamics*, W.R. Peltier, ed., Gordon and Breach Science Publ.

[Bus3] F.H. Busse, private communications 1978, 1989, 1995

[Bus] F.H. Busse: Lectures on fluid dynamics, unpublished

[BO] F.H. Busse, A.C. Or (1986) Subharmonic and asymmetric convection rolls. ZAMP **37**, 608–623

[BS] F.H. Busse, M. Sieber (1991): Regular and chaotic patterns of Rayleigh-Bénard convection. [SSKT, 79 – 92]

[Can] D.S. Cannell, private communication

[Cam] D.K. Campbell, ed. (1990): *Chaos/Xaoc, Soviet-American Perspectives on Nonlinear Science*, American Institute of Physics, New York,

[CB74] R.M. Clever, F.H. Busse (1974): Transition to time-dependent convection. J. Fluid Mech. **65**, 625–645

[CH] Courant, R.; Hilbert, D. (1924): *Methoden der Mathematischen Physik*, Bd. 1 und 2, (Berlin: Springer 1968, 3.Aufl.)

[Cha] S. Chandrasekhar: *Hydrodynamic and Hydromagnetic Stability* Oxford University Press, London 1961

[CoM92] K.T. Coughlin, P.S. Marcus (1992): Modulated waves in Taylor-Couette flow. J Fluid Mech **234**; Part 1: Analysis, p. 1–18; Part 2: Numerical simulation, p. 19–46

[CNJ] Chen, J.-C.; Neitzel, G.P.; Jankowski, D.F. (1987): Numerical experiments on the stability of unsteady circular Couette flow with random forcing. Phys. Fluids **30**, 1250–1258

[ChS] B. Chen, P. Saffman: Steady gravity-capillary waves on deep water I, II. Stud. Appl. Math. **60** (1979) 183–210; **62** (1980) 95–111

[CI94] P. Chossat, G. Iooss (1994): *The Couette-Taylor Problem.* Springer Verlag, Berlin

[Cli] K.A. Cliffe: Numerical calculations of two-cell and single cell Taylor flows. J Fluid Mech **135** (1983) 219–233

[ClM85] K.A. Cliffe, T. Mullin (1985): A numerical and experimental study of anomalous modes in the Taylor experiment. J. Fluid Mech. **153**, 243–258

[CGS] K.A. Cliffe, T.J. Garratt, A. Spence (1993): Eigenvalues of the discretized Navier Stokes equation with application to the detection of Hopf bifurcations, Advances in Computational Mathematics **1**, issues 3 & 4, 337–356

[Co65] D. Coles: Transition in circular Couette flow. J. Fluid Mech. 21(1965), 385–425

[DDS] A. Davey, R.C. DiPrima, J.T. Stuart (1968): On the instability of Taylor vortices, J. Fluid Mech. **31**, 17–52

[Da69] S.H. Davis (1969): On the principle of exchange of stabilities. Proc. Roy. Soc. A **310**, 341–358

[DK] N. Dinar, H.B. Keller: Computations of Taylor vortex flows using multigrid continuation methods. in: *Recent Advances in Computational Fluid Dynamics*, Proc. Princeton 1988, C.C. Chao, S.A. Orszag, W. Shyy (eds.), Lecture Notes in Engineering **43**, Springer Verlag Berlin 1989

[Doe] E. Doedel: AUTO, a package for the solution of continuation problems in ordinary differential equations. Appl. Math., Caltech 1985

[DPE] R.C. DiPrima, P.M. Eagles: Amplification rates and torques for Taylor vortex flows between rotating cylinders. Phys. Fluids **20** (1977), 171–175

[DES71] DiPrima, R. C.; Eckhaus, W.; Segel, L. A. (1971): Non-linear wave-number interaction in near-critical two-dimensional flows. J. Fluid Mech. **49**, pp. 705–744

[DPH] R.C. DiPrima, P. Hall (1984): Complex eigenvalues for the stability of Couette flow Proc. R. Soc. Lond. A **396**, 75–94

[DPR] DiPrima, R. C.; Rogers, E. H. (1969): Computing problems in nonlinear hydrodynamic stability, Phys. Fluids **12**, pp.: II-155–II-165

[DPS] R.C. DiPrima, H.L. Swinney: Instabilities and transition in flow between concentric rotating cylinders. In [SG, 139-180]

[DAC84] M.A. Dominguez-Lerma, G. Ahlers, D.S. Cannell (1984): Marginal stability curve and linear growth rate for rotating Couette-Taylor flow and Rayleigh-Rayleigh-Bénard convection, Phys. Fluids **27**, 856–860

[Do92] R.J. Donnelly (1992): Evolution of instrumentation for Taylor-Couette flow. [AH92, 1 - 27]

[DL87] R.J. Donnelly, M. M. LaMar (1987): Absolute measurement of the viscosity of classical and quantum fluids by rotating-cylinder viscometers. Phys. Rev. A **36**, 4507–4510

[DS] R.J. Donnelly, N.J. Simon: An empirical torque relation for supercritical flow between rotating cylinders. J. Fluid Mech. 7(1960), 401–418

[DR] P.G. Drazin, W.H. Reid: *Hydrodynamic Stability.* Cambridge University Press 1981

[Dy] *An Album of Fluid Motion.* M. van Dyke, ed. The Parabolic Press, Stanford 1982

[Eg71] P.M. Eagles (1971): On stability of Taylor vortices by 5th order amplitude expansions. J. Fluid Mech. **49**, 529–550

[Eg74] P.M. Eagles (1974): On the torque of wavy vortices. J. Fluid Mech. **62**, 1–9

[Eck65] W. Eckhaus (1965): Studies in Nonlinear Stability Theory. Springer Verlag, Berlin

[Eil] Eilbeck, J.C. (1987): Numerical study of bifurcation in a reaction diffusion model using pseudo-spectral and path following methods. Proc. Dortmund 1986, in: [KST], 47–59

[FBo] H. Fasel, O. Booz (1984): Numerical investigation of supercritical Taylor-vortex flow for a wide gap. J. Fluid Mech **138**, 21–52, see also [Boo]

[Fei93] M. Feistauer (1993): *Mathematical Methods in Fluid Dynamics.* Pitman, Longman Publ. Group

[FM] G. Frank, R. Meyer-Spasche: Computation of transitions in Taylor vortex flows. ZAMP **32** (1981) 710–720

[Fre] *Synergetics – From Microscopic to Macroscopic Order.* E. Frehland, ed. Springer Verlag, New York, Berlin, Heidelberg 1984

[GG93] T. Gebhardt, S. Großmann (1993): The Taylor-Couette eigenvalue problem with independently rotating cylinders. Z. Phys. B-Condensed Matter B **90**, 475–490

[GG94] T. Gebhardt, S. Großmann (1994): Chaos transition despite linear stability, Phys. Rev. E **50**, 3705–3711

[GO] D. Gottlieb, S.A. Orszag (1977): *Numerical Analysis of Spectral Methods: Theory and Applications.* SIAM, Philadelphia

[GH] J. Guckenheimer, P. Holmes (1983): *Nonlinear Oscillations, Dynamical Systems, and Bifurcations of Vector Fields.* New York,NY: Springer

[Ha80a] Hall, P. (1980): Centrifugal instabilities of circumferential flows in finite cylinders: nonlinear theory. Proc. Roy. Soc. Lond. A **372**, pp.: 317–56

[Ha80b] Hall, P. (1980): Centrifugal instabilities in finite containers: a periodic model. J. Fluid Mech. **99**, pp.: 575–96

[Hen] M.E. Henderson, unpublished results in connection with: M.E. Henderson, H.B. Keller: Complex bifurcations, to be published (1986). The computations were performed with an extension of John Bolstad's code, [BoKe].

[Hol83] M. Holderied (1983): Die Taylor-Rotation und ihre Anwendung zur Messung der Rotationsviskosität von magnetischen Kolloiden. Diplomarbeit in Physik (diploma thesis), U. München, 80 pages

[Hu92] J. Huitfeldt (1992): Prediction of critical boundaries in two-parameter nonlinear eigenvalue problems. Numerical Analysis Report **20**, Dept. of Computer Sci., Chalmers U. of Technology, Göteborg; 17 pages

[HR90] J. Huitfeldt; A. Ruhe (1990): A new algorithm for numerical path following applied to an example from hydrodynamical flow. SIAM J. Sci. Stat. Comput. **11**, pp. 1181–1192

[Jef28] H. Jeffreys (1928): Some cases of instability in fluid motion. Proc. Roy. Soc. A **118**, 195–208; Also Collected Papers (1975), **4**, 469–84

[Jo81] C.A. Jones: Nonlinear Taylor vortices and their stability. J. Fluid Mech. 102 (1981), 249–261

[Jo82] C.A. Jones (1982): On flow between counterrotating cylinders. J. Fluid Mech. 120, 433–450

[JM79] C.A. Jones, D.R. Moore (1979): Stability of axisymmetric convection. Geophys. Astrophys. Fluid Dyn. **11**, 245–270

[KB91] M. Kropp, F.H. Busse (1991): Thermal convection in differentially rotating systems. Geophys. Astrophys. Fluid Dynamics **61**, 127–148

[Ke77] H.B. Keller: Numerical solution of bifurcation and nonlinear eigenvalue problems. 359–384 in [Rab]

[Ke78] H.B. Keller(1978): Global homotopies and Newton methods. pp 73–94 in *Recent Advances in Numerical Analysis*, C. de Boor, G. Golub, eds., Academic Press, New York

[Ke87] H.B. Keller: *Lectures on Numerical Methods in Bifurcation Problems.* Published for the Tata Institute, Springer Verlag Berlin, Heidelberg, New York, Tokyo 1987

[KSc] H.B. Keller, W. Schröder (199.): Wavy Taylor-Vortex flows. In: *Computing Methods in Applied Sciences and Engineering*, R. Glowinski, A. Lichnewski, eds. SIAM, Philadelphia

[KS85] H.S. Kheshgi, L.E. Scriven (1985): Viscous flow through a rotating square channel. Phys Fluids **28**, 2968–2979

[Kir] K. Kirchgässner: Die Instabilität der Strömung zwischen zwei rotierenden Zylindern gegenüber Taylor-Wirbeln für beliebige Spaltbreiten. ZAMP 12(1961), 14–30

[KS68] K. Kirchgässner, P. Sorger (1968): Stability analysis of branching solutions of the Navier-Stokes equations. *Proc. 12th Intern. Congr. Appl. Mech.*, Stanford, pp. 257–268

[KS69] K. Kirchgässner, P. Sorger (1969): Branching analysis for the Taylor problem. Quart J Mech Appl Math **22**, 183–209

[KG83] E. Knobloch, J. Guckenheimer (1983): Convective transitions induced by a varying aspect ratio. Phys. Rev. A **27**, 408–417

[K91] E. Knobloch, private communication, 1991

[KD] S. Kogelman, R.C. DiPrima (1970): Stability of spatially periodic supercritical flows in hydrodynamics. Phys. Fluids **13**, 1–11

[KL89] H.-O. Kreiss, J. Lorenz (1989): *Initial-Boundary Value Problems and the Navier-Stokes Equations.* Academic Press, Boston

[KM83] M. Kubiček, M. Marek (1983): *Computational Methods in Bifurcation Theory and Dissipative Structures.* Springer Verlag, New York, Berlin, Heidelberg, Tokyo

[KST] Küpper, Seydel, Troger, eds. (1987) *Bifurcation: Analysis, Algorithms, Applications*, Proc. Dortmund 1986, ISNM 79, Birkhäuser Verlag

[Lad65] O. A. Ladyženskaya (1965): *Funktionalanalytische Untersuchungen der Navier-Stokes-schen Gleichungen.* Berlin

[Lad69] O. A. Ladyženskaya (1969): *The Mathematical Theory of Viscous Incompressible Flow*, Gordon & Breach

[Lam] H. Lamb (1932): *Hydrodynamics*, Reprint of 6th Edition, 1945; Dover Publications, New York
— (1931): *Lehrbuch der Hydrodynamik.* Autorisierte deutsche Ausgabe, mit Zusätzen von R. von Mises. Übersetzung der 5. engl. Auflage (1879, 1895, 1906, 1916, 1924); B.G. Teubner, Leipzig, Berlin

[LTK88] W.F. Langford, R. Tagg, E. Kostelich, H.L. Swinney, M. Golubitzky (1988): Primary instabilities and bicriticality in flow between counterrotating cylinders. Phys. Fluids **31**, 776–785

[Lo93a] D. Lortz(1993): *Hydrodynamik.* BI Wissenschaftsverlag. Mannheim

[Lo93b] D. Lortz (1993): On Rayleigh's stability criterion for Couette flow. Z. Naturforsch. **48a**, 703–704

[LMP] D. Lortz, R. Meyer-Spasche, P. Petroff (1991): A global analysis of secondary bifurcations in the Bénard problem and the relationship between the Bénard and Taylor problems. In: *Direct and Inverse Boundary Value Problems*, R. Kleinman, R. Kress, E. Martensen, eds., Methoden und Verfahren der mathematischen Physik **37**, Proc. Oberwolfach 1989, Verlag Peter Lang, Frankfurt a.M.,

[LM82] D. Lortz and R. Meyer-Spasche (1982): On the decay of symmetric dynamo fields. Math. Meth. Appl. Sci. **4** (1982), 91–97

[Ma81] Marcus, Ph. S. (1981): Effects of truncation in modal representations of thermal convection. J. Fluid Mech. **103**, 241–251

[Ma84] Marcus, Ph. S. (1984): Simulation of Taylor-Couette flow. J. Fluid Mech. **146**
I. Numerical methods and comparison with experiment, pp.: 45–64;
II. Numerical results for wavy-vortex flow with one travelling wave, pp.: 65–113

[Marg] Margules, Max (1881): Über die Bestimmung des Reibungs- und Gleitungs-Coefficienten aus den Bewegungen einer Flüssigkeit. Wiener Berichte (2. Folge) **83**, 588–602

[Mat90] M. Matern (1990): Einfluß von doppelten Eigenwerten auf sekundäre Verzweigungen in zähen Flüssigkeiten. Diplomarbeit in Theor. Physik, U. München, 65 p.

[Mey66] K.A. Meyer (1966): A 2-dimensional time-dependent numerical study of rotational Couette flow, Los Alamos Scientific Laboratory Report LA-3497

[Mey67] K.A. Meyer (1967): Time-dependent numerical study of Taylor-vortex flow. Phys Fluids **10**, 1874–1879

[Mey69] K.A. Meyer (1969): Three-dimensional study of flow between concentric cylinders. Phys Fluids **12**, Suppl. II (High Speed Computing in Fluid Dynamics), pp.: 165–170

[Me75] R. Meyer-Spasche: Numerische Behandlung von elliptischen Randwertproblemen mit mehreren Lösungen und von MHD-Gleichgewichtsproblemen. Ph.D. dissertation, Hamburg 1975; Report IPP 6/141; 92 p.

[Me82] R. Meyer-Spasche: Numerical approximation of boundary value problems having several solutions. Invited talk, In: *Metodi matematici nella dinamica dei fluidi e dei gas ionizzati* Proc. Seminario fisico matematico – Trieste, 28–29 Sept. 1981, 97ff; Trieste, 1982

[Me92] R. Meyer-Spasche (1992): Double Eigenvalues and the Formation of Flow Patterns. In: *Ordered and Turbulent Patterns in Taylor-Couette Flow*; Proc. NATO Advanced Research Workshop, Columbus, OH, May 22–24, 1991; C.David Andereck, Fernand Hayot, eds.; pp. 179–186, Plenum Press, New York

[MK80] R. Meyer-Spasche, H.B. Keller: Computations of the axisymmetric flow between rotating cylinders. J. Comp. Phys. 35(1980), 100–109

[MK78] R. Meyer-Spasche, H.B. Keller: Numerical study of Taylor vortex flows between rotating cylinders. Appl. Math. Report, Caltech July 1978, 64 p.

[MK85] R. Meyer-Spasche, H.B. Keller: Some bifurcation diagrams for Taylor vortex flows. Phys. Fluids **28** (1985), 1248–1252

[MK84] R. Meyer-Spasche, H.B. Keller: Numerical study of Taylor vortex flows between rotating cylinders II. Appl. Math. Report, Caltech March 1984, 38 p.

[MW1] R. Meyer-Spasche, M. Wagner: Steady axisymmetric Taylor vortex flows with free stagnation points of the poloidal flow. Proc. Dortmund 1986, in: [KST], 213–221

[MW2] R. Meyer-Spasche, M. Wagner: The basic (n, 2n)-fold of steady axisymmetric Taylor vortex flows. Proc. Tübingen 1986, *The Physics of Structure Formation*, W. Güttinger, G. Dangelmayr, eds., Springer Series in Synergetics, (1987), 166–178

[Mo78] H.K. Moffat (1978): *Magnetic Field Generation in Electrically Conducting Fluids*, Cambridge University Press

[Na86] M. Nagata (1986): Bifurcations in Couette flow between almost corotating cylinders. JFM **169**, 229–50

[Nea87] M.D. Neary, K.D. Stepanoff (1987) Shear-layer-driven transition in a rectangular cavity. Phys. Fluids **30**, 2936

[Nev87] M. Neveling, D. Lang, P. Haug, W. Güttinger, G. Dangelmayr (1987) Interaction of stationary modes in systems with 2 and 3 spatial degrees of freedom. Proc. Tübingen 1986, *The Physics of Structure Formation*, W. Güttinger, G. Dangelmayr, eds., Springer Series in Synergetics, (1987), 153–165

[Nei84] Neitzel, G.P. (1984): Numerical computation of time-dependent Taylor-vortex flows in finite-length geometries. J. Fluid Mech. **141**, pp.: 51–66

[ND81] Neitzel, G.P.; Davis, S.H. (1981): Centrifugal instabilities during spin-down to rest in finite cylinders; Numerical experiments. J. Fluid Mech. **102**, pp.: 329–52

[OvY] S.N. Ovchinnikova, V.I. Yudovich (1974): Stability and bifurcation of Couette flow in the case of a narrow gap between rotating cylinders. Prikl. Mat. Mekh. **38, no 6**

[Paf] M. Paffrath (1986) Das Taylorproblem und die numerische Behandlung von Verzweigungen. Ph. D. dissertation in mathematics, U. Bonn

[Pe88] P. Petroff (1988) Verzweigungsprobleme bei Strömungen in zähen Flüssigkeiten. Diplomarbeit in Theoretischer Physik, U. München, 58 p.

[Plu91] M. Plum (1991): Existence proofs in combination with error bounds for approximate solutions of weakly nonlinear second-order elliptic boundary value problems. ZAMM **71**, T660–T662

[Rab] *Applications of Bifurcation Theory.* P. Rabinowitz, ed., Academic Press, New York 1977

[Rie] H. Riecke (1986): Wellenlängeneinschränkung in quasi-eindimensionalen strukturbildenden Systemen. Ph. D. dissertation in physics, U. Bayreuth

[RP86] H. Riecke, H.-G. Paap (1986): Stability and wave vector restriction of axisymmetric Taylor vortex flows. Phys. Rev. A **33**, 547–553

[RP87] Riecke, H.; Paap, H.-G. (1987): Perfect wave-number selection and drifting patterns in ramped Taylor vortex flow. Phys. Rev. Lett. **59**, no.22, pp.: 2570–3

[RP90] H. Riecke, H.-G. Paap (1990): Wave-number restriction and mode interaction in Taylor vortex flow: appearance of a short-wavelength instability. Phys. Rev. A **41** 1943–1951

[Rieck] H. Riecke, private communications

[Rob65] P.H. Roberts (1965): The solution of the characteristic value problems (appendix to a paper by Donnelly and Schwarz). Proc. Roy. Soc. A **283**, 550

[RoB69] E.H. Rogers, D.W. Beard (1969): A numerical study of wide-gap Taylor vortices. J. Comp. Phys. **4**, 1–18

[Roe85] K. Roesner (1985): Ecoulement de Couette-Taylor stationnaire pour une solution de polymères. C.R. Acad. Sc. Paris, t. **301**, Série II, n° 20, 1387–1390

[Sa83] D.H. Sattinger (1983): *Branching in the presence of symmetry.* CBMS Regional Conference Series in Appl. Math., SIAM, Philadelphia

[Sae] D. Schaeffer (1980): Qualitative analysis of a model for boundary effects in the Taylor problem, Math. Proc. Camb. Phil. Soc. **87**, 307–337

[S90] J. Scheurle (1990): Analysis sukzessiver Instabilitäten am Beispiel des Couette-Taylorproblems. GAMM-Mitteilungen, Heft 2, p. 5–18

[SLB] A. Schlüter, D. Lortz, F. Busse (1965): On the stability of steady finite amplitude convection. J. Fluid Mech. **23**, 129–144

[Scr] Schrauf, G. (1986): The first instability in spherical Taylor-Couette flow. J. Fluid Mech. **166**, 287–303

[SK] Schreiber, I. and Kubiček, M., private communication, 1984

[ScK] Schröder, W., Keller, H. B. (1989): Wavy Taylor-vortex flows via multigrid-continuation methods, Caltech preprint, 1989 J. Comp. Phys.

[Sc86] J. Schröder (1986): Existence proofs for boundary value problems by numerical algorithms. Report Mathematisches Institut, U. Köln

[Seg62] L. A. Segel (1962): The non-linear interaction of two disturbances in the thermal convection problem. J. Fluid Mech. **14**, 97–114

[Sny] H.A. Snyder: Wave-number selection at finite amplitude in rotating Couette flow. J. Fluid Mech. 35, (1969) 273–298

[Strau] B. Straugham (1992): *The Energy Method, Stability, and Nonlinear Convection.* Springer Verlag, New York

[Stie] K. Stierstadt, U. München, private communication

[SG] *Hydrodynamic Instabilities and the Transition to Turbulence.* H.L. Swinney, J.P. Gollub (eds.), Springer Verlag, Berlin, Heidelberg, New York 1981, 1985 (2. ed.)

[SKC] H.L. Swinney, N. Kreisberg, W.D. McCormick, Z. Noszticius, G. Skinner (1990): Spatiotemporal patterns in reaction-diffusion systems, pp 197–204 in [Cam]

[SWM] H. Specht, M. Wagner, R. Meyer-Spasche: Interactions of secondary branches of Taylor vortex solutions. ZAMM **69** (1989), 339ff (FU-Preprint Nr. A-88-15, 1988)

[SW92] B.J. Schmitt, W. von Wahl (1992): Decomposition of solonoidal fields into poloidal fields, toroidal fields and the mean flow. Applications to the Boussinesq equations. In: *The Navier-Stokes Equations II – Theory and Numerical Methods*, Proc. Oberwolfach 1991, Lecture Notes in Mathematics **1530**, 291–305 Heywood, Masuda, Rautmann, Solonnikov (eds.), Springer Verlag Berlin

[SSKT] R. Seydel, F.W. Schneider, T. Küpper, H. Troger, eds.(1991): *Bifurcation and Chaos: Analysis, Algorithms, Applications.* Proc. Würzburg 1990, ISNM 97, Birkhäuser Verlag

[Stu] J.T. Stuart (1986): Taylor-vortex flow: a dynamical system. SIAM Rev. **28**, 315–342

[SD80] J.T. Stuart and R.C. DiPrima (1980): On the mathematics of Taylor vortex flows in cylinders of finite length, Proc. Roy. Soc. Lond. A **372**, 357–365

[TC] S.J. Tavener, K.A. Cliffe (1991): Primary flow exchange mechanisms in Taylor-Couette flow applying non-flux boundary conditions. IMA Journal on Applied Mathematics **46**, 165–199

[Tav] S.J. Tavener, private communication

[Tay] G.I. Taylor: Stability of a viscous liquid contained between two rotating cylinders. Phil. Trans. Roy. Soc. A 123(1923), 289–343

[Te83] R. Temam: *Navier-Stokes Equations and Nonlinear Functional Analysis.* SIAM, Philadelphia 1983

[Tew] R.P. Tewarson (1973): *Sparse Matrices.* Academic Press, New York

[TRD93] L.N. Trefethen, A.E. Trefethen, S.C. Reddy, T.A. Driscoll (1993): Hydrodynamic stability without eigenvalues. Science **261**, 578–584

[Tu86] L. Turyn (1986): Bifurcation without mixed mode solutions. In: *Multiparameter Bifurcation Theory*, M. Golubitsky, J.M. Guckenheimer, eds., Contemp. Math. **56**, AMS, Providence, 335–341

[Ve66] W. Velte: Stabilität und Verzweigung stationärer Lösungen der Navier-Stokes-schen Gleichungen beim Taylorproblem. Arch. Rat. Mech. Anal. 22 (1966), 1–21

[WS87] M. Wagner, H. Specht (1987): Numerische Studien zum Taylorproblem. Diplomarbeit in Mathematik (diploma thesis), FU Berlin, 289 pages

[Wer] B. Werner (1984): Regular systems for bifurcation points with underlying symmetries. Proc. Dortmund 1983, in [KMW]

[Wim] M. Wimmer: Viscous flows and instabilities near rotating bodies. Prog. Aerospace Sci **25** (1988) pp 43–103

[Yih] C.-S. Yih (1972): Spectral theory of Taylor vortices. Arch. Rat. Mech. Anal. **46**, 218–240 and **47**, 288–300

Index

Mathematics with Birkhäuser

International Series of Numerical Mathematics

Edited by
K.-H. Hoffmann, Technische Universität München, Germany
H.D. Mittelmann, Arizona State University, Tempe, USA
J. Todd, California Institute of Technology, Pasadena, USA

International Series of Numerical Mathematics is open to all aspects of numerical mathematics. Some of the topics of particular interest include free boundary value problems for differential equations, phase transitions, problems of optimal control and optimization, other nonlinear phenomena in analysis, nonlinear partial differential equations, efficient solution methods, bifurcation problems and approximation theory. When possible, the topic of each volume is discussed from three different angles, namely those of mathematical modeling, mathematical analysis, and numerical case studies.

ISNM 97 **R.U. Seydel, F.W. Schneider, T.G. Küpper, H. Troger (Ed.):** Bifurcation and Chaos: Analysis, Algorithms, Applications, 1991 (3-7643-2593-3)

ISNM 98 **W. Hackbusch, U. Trottenberg (Ed.):** Multigrid Methods III, 1991 (3-7643-2632-8)

ISNM 99 **P. Neittaanmäki (Ed.):** Numerical Methods for Free Boundary Problems, 1991 (3-7643-2641-7)

ISNM 100 **W. Desch, F. Kappel, K. Kunisch (Ed.):** Estimation and Control of Distributed Parameter Systems, 1991 (3-7643-2676-X)

ISNM 101 **G. Del Piero, F. Maceri (Ed.):** Unilateral Problems in Structural Analysis IV, 1991 (3-7643-2487-2)

ISNM 102 **U. Hornung, P. Kotelenez, G. Papanicolaou (Ed.):** Random Partial Differential Equations, 1991 (3-7643-2688-3)

ISNM 103 **W. Walter (Ed.):** General Inequalities 6, 1992 (3-7643-2737-5)

ISNM 104 **E. Allgower, K. Böhmer, M. Golubitsky (Ed.):** Bifurcation and Symmetry, 1992, (3-7643-2739-1)

ISNM 105 **D. Braess, L.L. Schumaker (Ed.):** Numerical Methods in Approximation Theory, Vol. 9, 1992 (3-7643-2746-4)

ISNM 106 **S.N. Antontsev, K.-H. Hoffmann, A.M. Khludnev (Ed.):** Free Boundary Problems in Continuum Mechanics, 1992 (3-7643-2784-7)

ISNM 107 **V. Barbu, F.J. Bonnans, D. Tiba (Ed.):** Optimization, Optimal Control and Partial Differential Equations, 1992 (3-7643-2788-X)

ISNM 108 **H. Antes, P.D. Panagiotopoulos:** The Boundary Integral Approach to Static and Dynamic Contact Problems. Equality and Inequality Methods, 1992 (3-7643-2592-5)

ISNM 109 **A.G. Kuz'min:** Non-Classicall Equations of Mixed Type and their Applications in Gas Dynamics, 1992 (3-7643-2573-9)

ISNM 110 **H.R.E.M. Hörnlein, K. Schittkowski (Ed.):** Software Systems for Structural Optimization, 1992 (3-7643-2836-3)

ISNM 111 **R. Burlisch, A. Miele, J. Stoer, K.H. Well:** Optimal Control, 1993 (3-7643-2887-8)

ISNM 112 **H. Braess, G. Hämmerlin (Ed.):** Numerical Integration IV. Proceedings of the Conference at the Mathematical Research Institute at Oberwolfach, November 8-14, 1992, 1993 (3-7643-2922-X)

ISNM 113 **L. Quartapelle:** Numerical Solution of the Incompressible Navier-Stokes Equations, 1993 (3-7643-2935-1)

ISNM 114 **J. Douglas, U. Hornung (Ed.):** Flow in Porous Media. Proceedings of the Oberwolfach Conference, June 21-27, 1992, 1993 (3-7643-2949-1)

ISNM 115 **R. Bulirsch, D. Kraft (Ed.):** Computational Optimal Control, 1994 (3-7643-5015-6)

ISNM 116 **P.W. Hemker, P. Wesseling (Ed.):** Multigrid Methods IV. Proceedings of the Fourth European Multigrid Conference, Amsterdam, July 6-9, 1993, 1994 (3-7643-5030-X)

ISNM 117 **R.E. Bank, R. Bulirsch, H. Gajewski, K. Merten (Ed.):** Mathematical Modelling and Simulation of Electrical Circuits and Semiconductor Devices, 1994 (3-7643-5053-9)

ISNM 118 **W. Desch, F. Kappel, K. Kunisch (Ed.):** Control and Estimation of Distributed Parameter Systems: Nonlinear Phenomena. International Conference on Control and Estimation of Distributed Parameter Systems, Vorau, July 18-24, 1993, 1994 (3-7643-5098-9)

ISNM 119 **R. Zahar (Ed.):** Approximation and Computation: A Festschrift in Honor of Walter Gautschi, 1995 (3-7643-3753-2)

ISNM 120 **W. Hackbusch:** Integral Equations: Theory and Numerical Treatment, 1995 (3-7643-2871-1)

ISNM 121 **R. Jeltsch, M. Mansour (Ed.):** Stability Theory. Hurwitz Centenary Conference, Ascona, 1995, 1996 (3-7643-5474-7)

ISNM 122 **A.M. Khludnev, J. Sokolowski:** Modelling and Control in Solid Mechanics, 1997 (3-7643-5238-8)

ISNM 123 **C. Bandle, W.N. Everitt, L. Losonczi, W. Walter (Ed.):** General Inequalities 7. 7th International Conference, Oberwolfach, November 13–18, 1995, 1997 (3-7643-5722-3)

ISNM 124 **W.H. Schmidt, W.H. / K. Heier / L. Bittner (Ed.):** Variational Calculus, Optimal Control and Applications International Conference in Honour of L. Bittner and R. Klötzler, Trassenheide, Germany, September 23-27, 1996, 1998 (3-7643-5906-4)

ISNM 125 **G. Nürnberger / G. Walz / J.W. Schmidt (Ed.):** Multivariate Approximation and Splines, 1997 (ISBN 3-7643-5654-5)

ISNM 126 **W. Desch / F. Kappel / K. Kunisch (Ed.):** Control and Estimation of Distributed Parameter Systems. International Conference in Vorau (Austria), July 14–20, 1996, 1998 (3-7643-5835-1)

ISNM 127 **H.H. Gerke / U. Hornung / Y. Kelanemer / M. Slodicka / S. Schumacher:** Optimal Control of Soil Venting: Mathematical Modeling and Applications, 1999 (3-7643-6041-0)

GPSR Compliance

The European Union's (EU) General Product Safety Regulation (GPSR) is a set of rules that requires consumer products to be safe and our obligations to ensure this.

If you have any concerns about our products, you can contact us on ProductSafety@springernature.com

In case Publisher is established outside the EU, the EU authorized representative is:

Springer Nature Customer Service Center GmbH
Europaplatz 3
69115 Heidelberg, Germany

Batch number: 10584556

Printed by Printforce, the Netherlands